GEMS OF WASHINGTON

By
Lanny R. Ream

Introduction by Susan Cedarleaf

Jackson Mountain Press
P.O. Box 2652
Renton, WA 98056

Copyright © 1994
ISBN 0-918499-09-7

Distributed by:
Gem Guides Book Co.
315 Cloverleaf Drive, Suite F
Baldwin Park, CA 91706

Dedicated to my parents

The author would like to thank the many collectors, gem and mineral clubs and dealers who furnished information and specimens for photographs. This book is for them, but it was their help that made it possible. Special thanks go to Susan Cedarleaf for writing the introduction and for her many hours spent reviewing and assisting with the manuscript.

The author stuuying tremolite, Illinois Creek.

INTRODUCTION

The state of Washington has been attracting gem and mineral collectors for many years. In addition to an abundance of material, the collector is offered varied geology and geography which, together, create some of the world's most spectacular scenery.

The wide variety of gems and minerals occurring in Washington is due to the complex geology of the state. Of great importance to gem collectors are the Miocene age Columbia River basalts that cover the southeast, and the Eocene age basalt and andesite flows that cover the west. These volcanic rocks contain numerous deposits of agate, opal, petrified wood, and geodes.

Quartz, pyrite, autunite and beryl are but a few of the minerals found in vugs and cavities in the intrusive rocks, outcrops of which can be found in northeastern and northcentral Washington, as well as throughout the Cascade Mountains. Crystals are found lining vugs in the main plutons and in the many dikes that intrude the surrounding sedimentary and metamorphic rocks of the eastern, northcentral and northwestern parts of the state and the Cascade Mountains.

Washington's topography and climate is as varied as its geology. Collectors can choose from semi-arid to rainforest, and from smooth ocean beaches to rugged mountainsides. The low precipitation and dry, hot climate of the eastern part of the state have combined to create almost ideal conditions for discovering gem material. Very little soil has developed over the Columbia River basalts, hence vegetation is sparse and outcrops of gem-bearing rocks are often exposed. Thus, removal of petrified wood is much easier. However, the rockhound working under the withering summer sun, in temperatures of 95° and higher, may not fully appreciate this.

Western Washington receives more than its share of the state's annual rainfall, with some areas receiving more than 100 inches. This is conducive to heavy growth of vegetation and thick soil which cover many interesting deposits.

The Cascade Mountains have a totally different environment from the rest of the state. During the fall, winter, and spring months, collecting in the higher regions is almost impossible, as 20 to 80 feet of snow blankets the ground. During summer, when the alpine beauty is in its finest splendor, field trips into these mountainous areas can be very rewarding, both in collecting and enjoying the panorama. Collecting conditions here will consist of long, rugged hikes, climbing through narrow, sheer canyons, and hanging from steep cliffs. At least there are abundant rock exposures for the intrigued!

The eastern two-thirds of the state is composed of a semi-arid plateau, cut by rivers into deep wide canyons. The northwest is dominated by several ranges of low mountains; this area includes some of the most complex geology in the state, as well as most of its mineral deposits of present-day economic importance.

Scattered throughout the state are well-known gem and mineral deposits which have produced for years. Many additional deposits have been recently discovered which will provide collecting opportunities for many years to come. As collectors venture further from the easily accessible areas, the possibility that new important discoveries will be made is very high.

Land ownership within the state is varied. Much of the land (approximately 30%) is in the public domain, managed by the US Forest Service, Dept. of Fish and Game, and the Bureau of Land Management. On the east side of the state, most of the popular deposits are located on public land. On the west side are vast holdings of forested land owned by timber companies. These companies have been very cooperative in opening their lands to collectors. Some private land is also open to collectors. It is imperative that all collectors work to keep these lands open. Littering, vandalism, and over-collecting, done by just a few, can lead to the closing of even public lands to all collectors.

Information for this book was gathered from a variety of sources. Much of the data was obtained by the author's field trips to the deposits. Additional information was obtained from other collectors and publications.

Gem and mineral hobbyists who have collected in Washington agree that it is indeed, a collectors' paradise. The geological processes that have produced a wide variety of gems and minerals, have helped shape the diverse and beautiful panorama of magnificent scenery. There are hundreds of known deposits to collect in, and many new ones to discover. The gems, minerals and breathtaking beauty of this state are waiting to be collected and enjoyed by all.

Susan Cedarleaf

Collecting on state and federal lands is becoming increasingly regulated. Before collecting on ANY federal land, contact the appropriate USFS ranger district to find out if the lands are covered by mining claims; withdrawal areas; or are environmentally sensitive. Wilderness areas have special rules, and collecting needs to be authorized *in advance* by permit.

The state of Washington issues Rockhounding Permits for some areas: Walker Valley, Mt. Teneriffe, Tiger Mountain, and Saddle Mountain at present. This policy may be expanded to all state lands in the near future. For further information, contact the Dept. of Natural Resources in Olympia.

All photos by the author and specimens from the author's collection, except as otherwise noted.

PART I: GEMS

AGATE AND OTHER VARIETIES OF CRYPTOCRYSTALLINE QUARTZ

The cryptocrystalline variety of quartz is very common and is given the general name of chalcedony. It is not found as crystals, but is always found as masses, nodules, vein fillings or pseudomorphs of other minerals, plants or animals. The composition of chalcedony is a compact or felted mass of microcrystalline rods of quartz, which gives it a high degree of toughness.

Chalcedony comes in a variety of colors that form many patterns. The names of the different cryptocrystalline gems are based on the different colors and patterns. It is rarely colorless, but is commonly blue, gray, green, yellow, red or combinations of these colors.

Agate is the most common of the chalcedony gems. The term agate originally was used for only translucent material displaying color banding. In common usage the name is applied to other patterns. Moss agate is a type of chalcedony that contains dendrites that appear as moss-like inclusions. This type often shows very natural-looking scenes of trees and shrubs formed by the dendrites. Fortification agate is a type in which the alternating color bands are angular. If the color bands are not angular then they are called banded agates. There are other varieties of agate that are self explanatory, including breccia agate, and eye agate. Yellowish-red to red chalcedony is known as carnelian, and often shows banding. Dark red fracture-free material is rare and much sought after.

Some myrikite has been found in Washington. This is a type of chalcedony that contains inclusions of cinnabar. The red streaks in the clear chalcedony make a very interesting gem. One of the most popular and rare agates in the state is the type that is known as Ellensburg Blue. This banded blue agate is found only in an area north of Ellensburg.

Opaque types of cryptocrystalline quartz are quite common and are found in almost as many variations as the translucent types. Jasper is probably the most common form. The name is applied to all colors of opaque cryptocrystalline quartz, but was originally reserved for the red variety. For other colors and patterns the word jasper is commonly preceeded by a modifier, either the name of the locality or a descriptive term. A case in point is orbicular japser, which is composed of spheroids of red or other colored jasper in a matrix of red, green, gray or translucent chalcedony.

Chalcedony is a common pseudomorph of other minerals and organic matter. Petrified wood is often formed by wood tissue replaced by chalcedony. The wood may be replaced by chalcedony ion-by-ion, which perfectly preserves the cell structure, or the small voids between cell walls of the wood may be replaced partially preserving the original plant structure. In other cases the chalcedony has filled the mold formed in the enclosing rock by removal of the organic matter. In this case, only a cast of the wood remains. There is no internal structure of the wood in chalcedony.

Cryptocrystalline quartz is widespread in the state. It is most common in the basalts of eastern and western Washington as petrified wood, agate and jasper. Many metal mines contain chalcedony as a gangue mineral, usually white or gray but rarely of gem quality. Most of the river, ocean beach and glacial gravels contain pebbles of many types of chalcedony. The localities are described in the following pages

Ellensburg-Cle Elum Area

A large area extending north of a line drawn from Cle Elum to Ellensburg produces many types of agate. This area produces the state's most rare agate, Ellensburg Blue. This agate varies from grayish-blue to dark blue and is usually layered or banded. Most pieces of the agate are less than .5 pound in weight but some have been found that weigh more than 2 pounds. This agate takes an excellent polish, but care must be taken to expose the best blue layer.

There is controversy over just what Ellensburg Blue is. Some collectors maintain that only the darkest blue agates are Ellensburg Blue; other collectors say that all blue agate in this area is Ellensburg Blue. What Ellensburg Blue is may best be left up to the individual collector.

The best localities to find this material are the canyons of Dry Creek, Green Creek, Reecer Creek and Horse Canyon. All of these creeks are located northwest of Ellensburg. Highway 97 that goes north out of Ellensburg, follows Horse Canyon. The other canyons are shown on figure 1. Searching the ground and talus in the canyons and dry creek bottoms may be productive, but these areas are heavily prospected each year. Winter frost heaving brings new material to the surface, and many hours of careful searching will find it.

The low rounded ridges between these canyons are almost bare, covered only by clumps of short brown grass. Numerous pieces of light gray to blue agate have been found on these ridges. Most of these chips and pieces weigh less than an ounce and are less than 1.5 inches long. Most are layered, and a few are dark blue. A few pieces with mossy patterns, and pieces of carnelian and blood red jasper can be found.

Most collectors hunt for this agate by slowly walking over the ridges and up the canyons. In the last few years some good Ellensburg Blue has been found in the power line right of way, and in road cuts. A few collectors carry hammers and break chunks of the basalt searching for the elusive Ellensburg Blue.

Much of this land is public land administered by the Bureau of Land Management. Some of it is private, and most of it is grazing land for cattle. The public land is open to collecting. Some of the private land owners will allow rockhounds to collect on their property, if asked.

This area in Central Washington is very hot and dry during the summer. There is no water available for drinking, and the streams only flow during early spring runoff. All collectors should be prepared for this and carry an ample supply of water. The irrigation canal that flows across the lower portion of this area may look cool and refreshing, but its water is not suitable for human consumption.

There are camping areas nearby along the Cle Elum and Teanaway Rivers and nearby Swauk Creek. Collectors who are interested in Ellensburg, may wish to go to this area to collect in some of the other nearby localities.

The Teanaway basalt that contains the agates can be found in a large area that extends north and east from Easton around Cle Elum Lake to Yellow Hill, northwest of Casland, and north and south of Red Top Mountain. There is a great variety of agate in this basalt. The Ellensburg Blue agate is found mostly in the Ellensburg formation, which is alluvium that was formed by the erosion of the Teanaway basalt. Teanaway basalt can be distinguished from the younger Columbia River basalt by its reddish color and vesicular and scoriaceous flow tops. The Columbia River basalt is mostly dark gray to black and produces few agates in this area.

Williams and Boulder Creeks and Crystal Mountain east of Liberty contain some blue agate and quartz-lined geodes. This area is a .5 hour drive north of Ellensburg via Highways 131 and 97. A side road is taken to the old gold mining town of Liberty, and old roads can be followed to the various creeks in the area. These creeks flow from Table Mountain (Fig. 1), which is capped by Columbia River basalt. The agate-bearing Teanaway basalt is exposed below this cap. Crystal Mountain is west of Table Mountain. Nodules that have eroded out of the basalt can be found in the creek bottoms and canyon walls. Numerous agates have been recovered by breaking the basalt or by digging in the rocky soil. Rockhounds can bring the tools they prefer to use most: pick and shovel to dig with or hammer and gad to break rock. Much of this land is held under gold mining claims; rockhounds should obey all no trespassing signs and stay away from open mines.

By following the Robinson Gulch road (not passable by passenger car) for 2 miles then walking the trail for another .5 mile, the Crystal Mountain diggings can be reached. Surface material can be found along the trail to the digging area, which is on the left near the top of the mountain. More blue agate, crystal-lined geodes and jasper can be found here. Collectors will need mattocks, picks, bars, or shovels, in addition to hammers and gads to recover this material. It is found in solid rock and rocky soil. The digging conditions are very similar to the more famous Red Top Mountain diggings that are located northwest of here.

Red Top Mountain is the most popular collecting location in this area. It is accessible by trail and a gravel road that turns left off Highway 97 just north of the Mineral Springs Resort. This is the Blue Creek Road; follow it to the parking lot at the end of the road. The trail is taken from the parking lot towards the lookout but do not take the left fork to the lookout. The main agate beds are just north of the fork to the lookout, about a 1.5 mile hike, but can be found along the road to the south of the lookout too. The road is easily maneuvered by the average sedan, and the trail is wide and not steep.

All collectors will want to pack in a hammer and pick, mattock or bar. Sometimes a shovel will be useful, but most of the collecting is done in basalt or a very rocky soil. Be sure and pack in a lunch and water; a pack sack is needed to pack out specimens. Most of the collecting is hard work, but some surface material is still being found by collectors who hunt away from the main diggings.

Collecting is done in a large area on the ridge. The collector can choose from digging in the rocky soil, talus or the basalt and scoria. The material found is blue-gray to blue agate nodules, to 4 inches, crystallined agate geodes to 10 inches, thundereggs to 4 inches, and rare amethyst geodes. Geodes are found mostly on the extreme north and south ends of the ridge. Most of the ridge produces blue agate nodules. Small thundereggs can be found near the north end of the ridge.

Some of the areas to the south and west of Red Top Mountain have produced agates and geodes of similar material. Amethyst geodes have been reported from this area and in the area extending from the Middle Fork of the Teanaway River to Cle Elum Lake. Collectors will find that the areas near the roads are heavily worked. Much of the land between the known areas produce agates, but have not been worked by many collectors because of the difficult access. Ambitious rockhounds have found excellent quality agate nodules and geodes by hiking short distances from the roads.

Carnelian has been found at Yellow Hill on the Middle Fork of the Teanaway River. It is reached by taking the Teanaway Road about 7 miles east of Cle Elum on the Blewett Highway (Highway 97). At Casland turn right on the Middle Fork Road for about 4 miles to road 2110. This road is followed to a point .1 mile beyond the National Forest Boundary marker. From here it is necessary to walk along an old road to the digging area. Agates can be found in bulldozer cuts and the numerous pits that have been dug by energetic collectors. This area has become popular in recent years. Other canyons and hillsides in the vicinity produce similar agate and carnelian, but material is not adundant as at Red Top or Crystal Mountain.

Agate nodules and geodes similar to those of Red Top Mountain can be found on Easton Ridge, also known as Cle Elum Ridge. Digging conditions are similar to Red Top Mountain, but there is more soil here. This area is reached by taking the Roslyn exit from Interstate 90, 3.5 miles west of Cle Elum. Turn left off the overpass on the first logging road and follow it to the crest of the ridge. Most of the agate is found on the sides of the ridge to the south and east of the beacon; it is not as abundant as the material at Red Top.

Geodes and agates can be found on Frost Mountain 20 miles west of Thorp. This collecting area is reached by taking the Bruton Road just south of Thorp to Taneum Creek Road. Follow this road to the Forest Service Guard Station where Quartz Mountain Road begins. This road is followed to the meadow, park here. From the parking place it is necessary to hike up a trail to the lookout. Continue over the hill to the southeast side to where the diggings are located (Fig. 2). Calcite and quartz lined geodes are found in the basalts below the end of the bluff. Good material, including blue-gray agate, can be found in the talus. A hammer and chisel will be useful to work material out of the basalt.

The Cle Elum and Ellensburg Chambers of Commerce are excellent sources of information on the above collecting localities. Both Chambers of Commerce will mail pamphlets to rockhounds or provide firsthand information when visited. It is a good policy to visit these sources of information as well as the local clubs to obtain information on present digging conditions in this area.

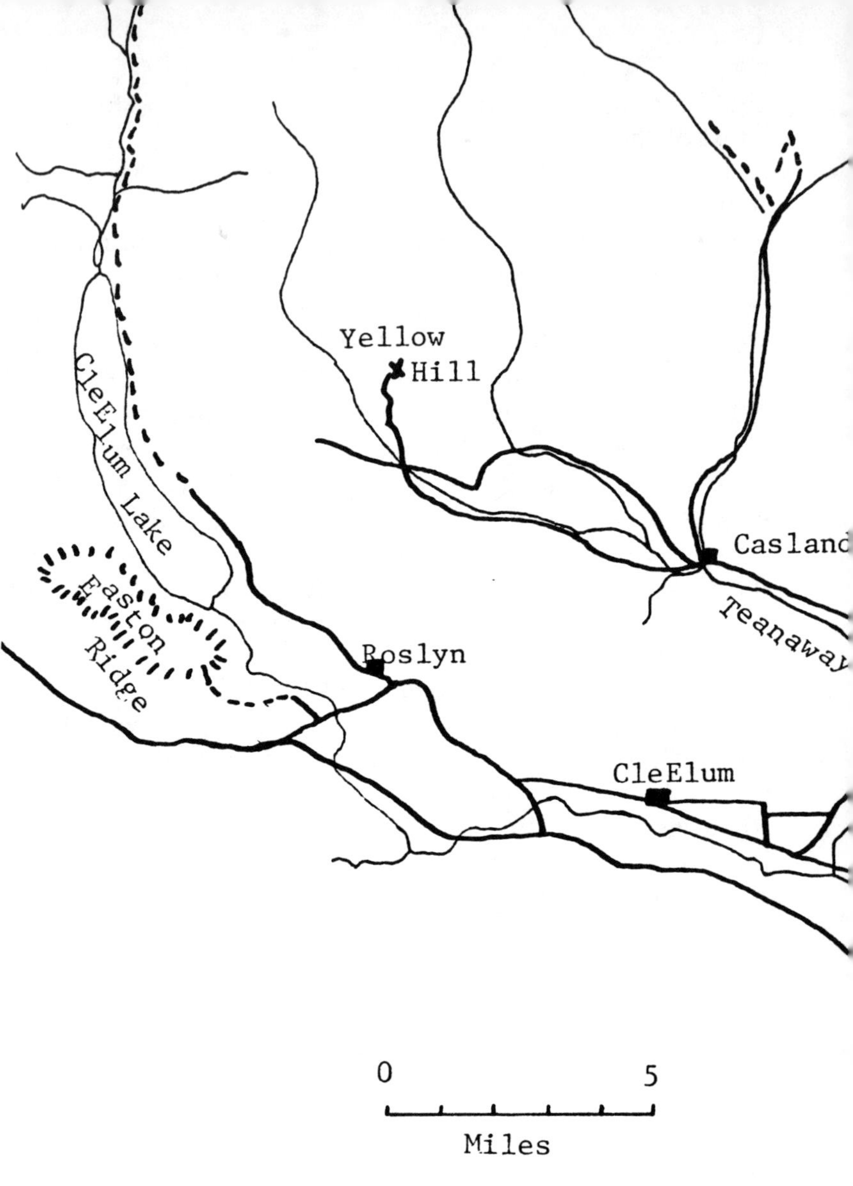

Fig. 1 *Ellensburg-Cle Elum area*

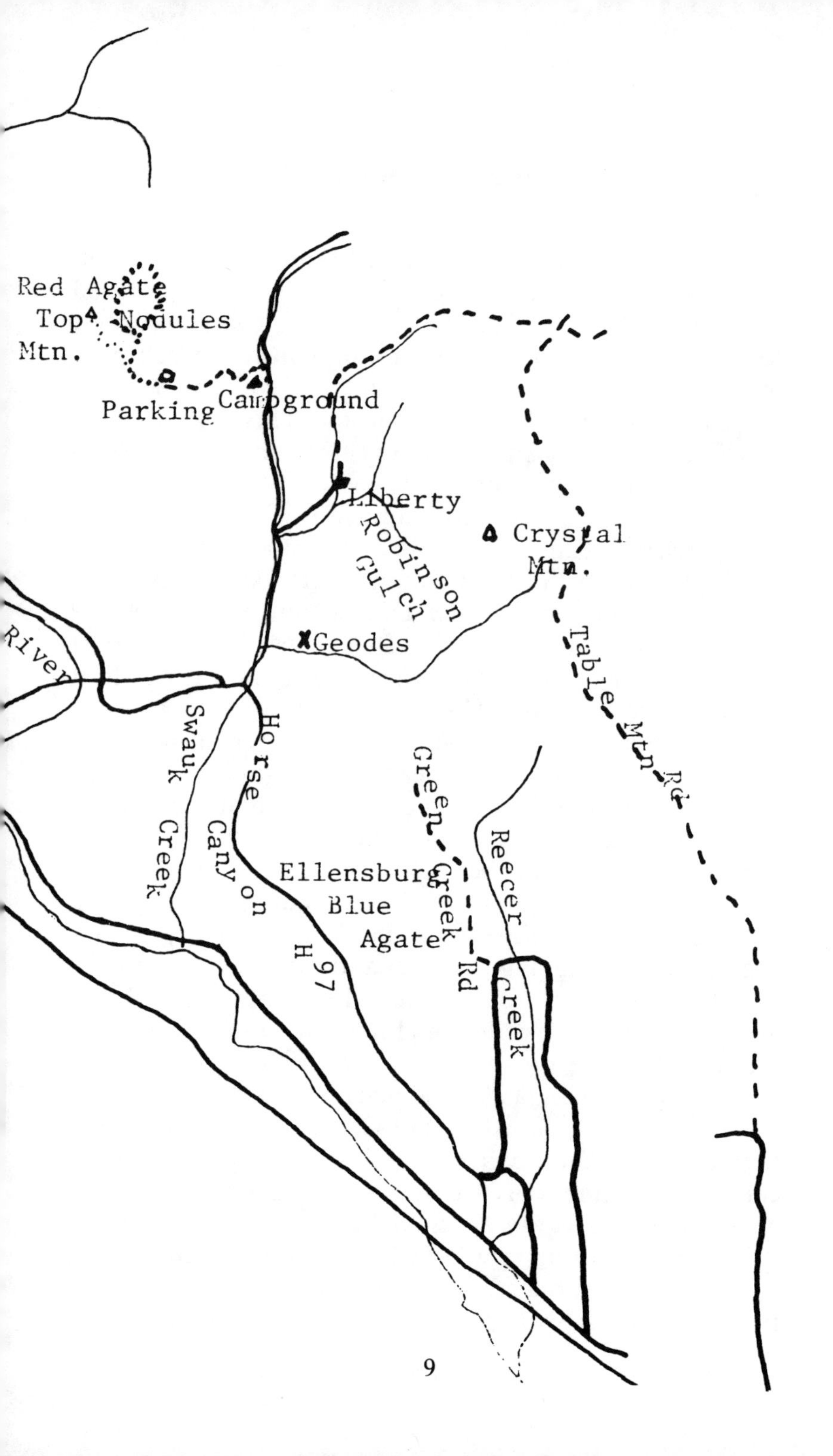

Red Agate
Top Nodules
Mtn.
Parking
Campground
Liberty
Robinson Gulch
Crystal Mtn.
Geodes
River
Swauk Creek
Horse Canyon
H 97
Ellensburg Blue Agate
Green Creek Rd
Reecer Creek
Table Mtn Rd
9

Vantage to Goldendale

In the eastern part of Kittitas County along the Columbia River there are scattered occurrences of petrified wood. Few of these areas are individually described here, because opalized wood is most common. A search near the opalized wood areas will provide the collector with an occasional piece of "agatized" wood. This area and the petrified wood deposits are more thoroughly described in the following chapter on opal.

The gravels of the Columbia River from Vantage downstream to Ringold contain agate and petrified wood. The high water behind Wanapum and Priest Rapids Dams covers most of the gravels. During unusual low water periods in the fall, gravels are often exposed. Below Priest Rapids the river is at its natural level, unfortunately it passes through the Hanford Atomic Reservation. Rockhounds cannot gain access to the river in the reservation. The stretch of the river from Ringold to Richland may be productive. This is the only stretch of the Columbia River that has not been dammed.

Petrified wood can be dug at Puckerhuddle, 2 miles northwest of White Salmon. The wood is found on a plateau near the confluence of the White Salmon and Columbia Rivers. Digging in basalt and rocky soil is required to collect this wood. Tools needed include a bar, shovel and hammer; some collectors find that a pick or mattock are useful in the rocky soil.

Agate, carnelian, jasper and petrified wood can be found in an area 2 miles northeast of Goldendale. This type of material and good opalized wood is found in many of the canyons. Digging tools are the same as those required for all the basalt digging areas in Eastern Washington.

The area around Table Mountain west of Stevenson (Fig. 3), has produced agate, carnelian and jasper. On the East Fork of Spring Creek, gray chalcedony cavity fill-

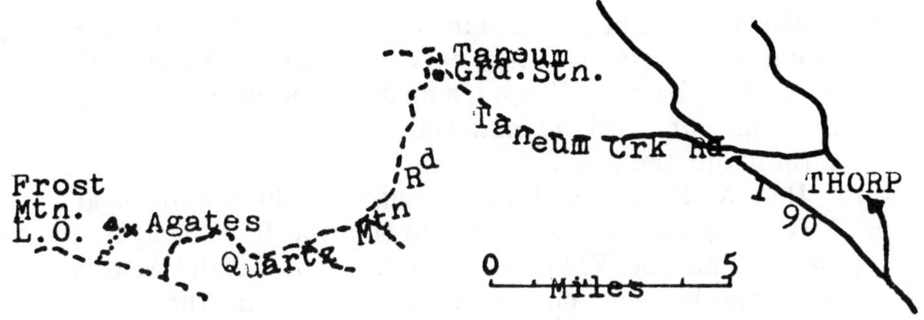

Fig. 2 *Frost Mountain area*

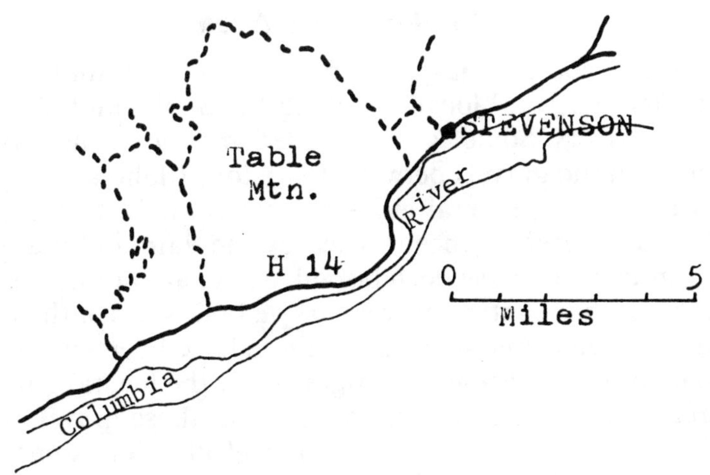

Fig. 3 *Table Mountain-Stevenson area*

ings in basalt have been found. These are in large pieces, up to 10 inches in length. This entire area is typical of the localities in the Columbia River Basalts; pieces of chalcedony can be found in most talus slopes and outcrops of basalt. Digging requirements are generally the same in most of the basalts. This is not an area of large high quality deposits but excellent material can be found scattered throughout the area.

Red and maroon jasper occurs in the basalt near Yakima along the east side of old Highway 97 halfway between Selah and Yakima near the exit to Selah. At this point the highway curves to the west around the ridge. Similar material may be exposed in the roadcuts of Highway 82 nearby, and some pieces of jasper have been found in the gravel bars of the Yakima River here.

The south slopes of the Saddle Mountains, on the east end, are known as the Wahluke Slopes. A recent discovery of picture jasper has been made in this area, and the material has been named Wahluke Jasper. It displays complex patterns of various shades of brown that makes pictures not unlike the dry, desert-like country where it is found. Presently it is being commercially produced and is not available for collecting by rockhounds.

Snake River Area

Agate nodules, jasper and opal can be found in the basalts in several localities along the Snake and Grande Ronde rivers. Some agates, banded and mossy, have been found in the gravels between Lewiston, Idaho and Lower Granite Dam, several miles downstream. Unfortunately the water level has been raised by the dam and many of the gravel bars have been buried. There are numerous old terraces that contain gravel bars on both sides of the river through this stretch. Some of these have been cut by the road and the railroad. See figure 4 for the location of this area. The author has found agates in these gravel bars, some of them are mossy and pastel in color, some are banded, and one small piece of sagenite was found.

The canyon walls are entirely Columbia River basalts, except for Granite Point. These layers of basalt are well illustrated by figure 5. This picture was taken looking downstream from a point near Moses. Moses was an old train stop where the river makes a sharp bend and changes from its westerly flow from Clarkston to a northerly direction.

Small chalcedony vesicle fillings and pieces of jasper can be found in this area. Large nodules of green, black and creamy colored chalcedony (Fig. 6) can be found in the canyon near Moses. Figure 4 gives the location of gem material in this area. The canyon that contains the jasper also contains the Moses precious opal deposit. See the following chapter for a description of this material and a description of access to these two gem materials.

The chalcedony nodules have to be dug out of solid basalt pillows. Collectors have worked the material out of the basalt in holes that extend arms length into the basalt. The nodules are 3 to 4 inches in diameter and over 10 inches in length. Most of the material contains finely disseminated pyrite that gives it an interesting sparkle. It takes an excellent polish. Tools for collecting in hard rock are needed at this locality.

This is an excellent area for wandering and prospecting for chalcedony gems. Small pieces of chalcedony are abundant as small vesicle fillings (amygdules) and as chips. They can be found in many colors including white, gray, green, red and colorless. Some have small quartz crystals in the center. Small blue-gray and white opal amygdules can be found too, by wandering over the talus and basalt cliffs. This is a pleasant way to spend an afternoon and collect material that is great for tumbling.

Agate nodules and pebbles are reported from the gravels at the mouth of the Grande Ronde River and in gravel bars upstream. Some of the agate is reported to be moss agate, and pieces of petrified wood have been found. The Grande Ronde River passes through many layers of Columbia River basalts (Fig. 7), so there is ample host rock

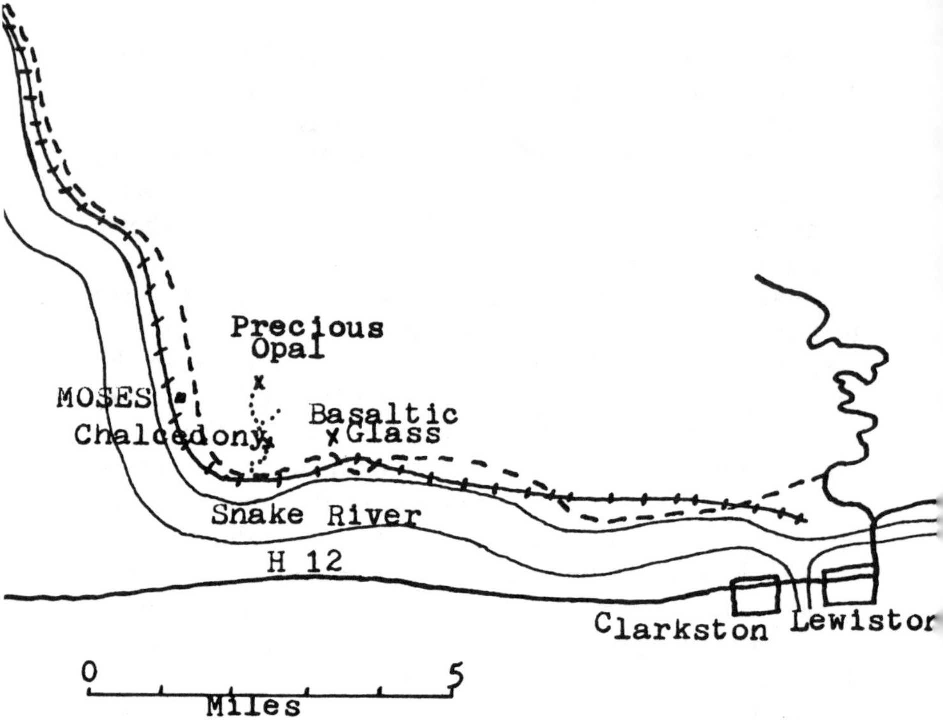

Fig. 4 *Snake River canyon area*

Fig. 5 *Snake River canyon below Moses. The river has cut through numerous basalt flows, some of which contain gem material.*

14

to provide good gem material to the numerous gravel bars. The canyon is very narrow and meandering; access is very limited.

About 4 miles up Asotin Creek from Asotin, south of Clarkston, is a small deposit of petrified wood. The deposit is on the right side of the road and creek in the hillside. There is a parking area along the road. Only a short walk is required to reach the deposit. The wood is very solid and brown in color. Most of it shows the grain very well, in fact much of it looks like real mahogany-stained wood. Logs up to 18 inches in diameter and over 15 feet in length have been removed from this deposit.

The host rock is a pillow basalt. A collector will have to plan on much hard work to collect excellent petrified wood here. Other small deposits of petrified wood have been found nearby, most of these are on private land. Search the creek bottoms for pieces of wood and scour the hillsides for outcrops of pillow basalt. This area is well worth prospecting.

Northeastern Washington

Very little gem material has been reported from this portion of the state, as few outcrops of volcanic rocks are exposed.

Near Davenport in Secs. 7 and 8, T.22N., R.17E., on the north side of Negro Creek, agate-lined geodes have been reported. These are found in the basalt, or weathered out onto the surface. Most of this is private land; obtain permission before collecting.

Agate and quartz geodes have been reported from an area north of Curlew and from an area 15 miles northeast of Republic. A few amethyst geodes to over 2 feet in diameter have been collected at Fransom Park near Curlew.

Most of the rocks here are granitic and metamorphic, but there are Eocene to Oligocene age volcanics in two areas, one of which extends northeast and southwest from Republic for several miles, the other is 10 to 15 miles west

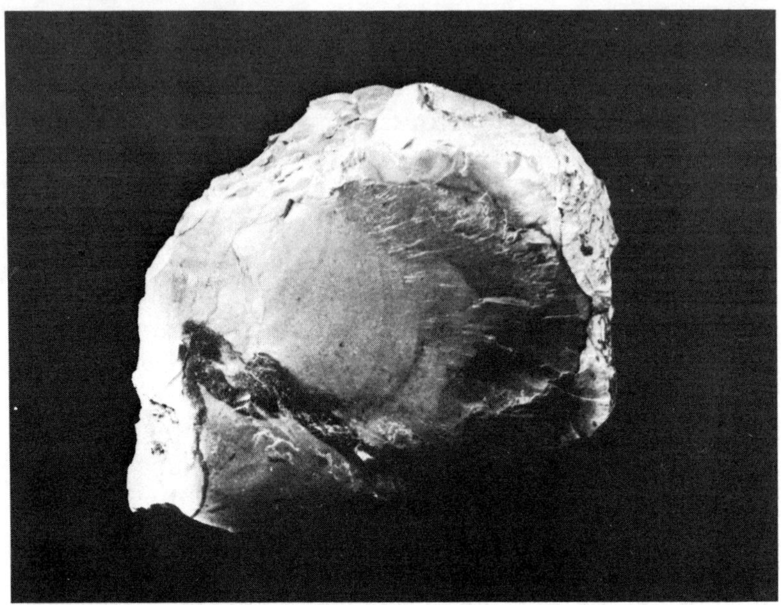

Fig. 6 *Green and black chalcedony from Moses.*

Fig. 7 *Grande Ronde River Canyon. Note the basalt dike, center, which is believed to be a source of the flows.*

of Curlew. Small outcrops can be found scattered through much of this region.

These volcanics are andesites, tuffs and basalts. They are good hosts for agate nodules and geodes. It is possible that there is some petrified wood in a few locations. This region has not been very well explored by rockhounds, but has good potential.

Blue agate occurs in seams and veins in basalt southeast of Tonasket at Frosty Creek. The agate varies from white to blue and most show fortification banding. Drusy quartz crystals cover some of it. Hard rock tools are needed to break the basalt to expose the agate.

To reach this locality take Highway 20 east out of Tonasket for about 17 miles to the Aeneas Valley Road. Turn right on this road and follow it for 15.6 miles to the Frosty Creek Road. Turn left on this road, follow it for 1 mile, turn right and follow this road for .3 mile. At this fork take the right branch and follow it for 1.1 miles to where the road crosses a wash. The agate is exposed on the right (south) side of the road on the east side of the creek, in the cliff.

Southwestern Washington

Many types of chalcedony, including carnelian, banded agate, quartz-filled geodes and jasper, can be found in the Lucas Creek, Newaukum River and Pigeon Springs areas. The material is found as nodules and pieces up to several pounds in weight. An ocasional piece of petrified wood may be found, but it is not abundant.

The source of the gem material is the plentiful volcanic rocks of Eocene age. These rocks, consisting of andesite, basalt, breccias and pyroclastics, were deposited in a eugeosynclinal trough that occupied the present area of western Washington in early Tertiary times (Snavely and Wagner, 1963). These volcanics were most erupted onto the floor of the shallow trough. Younger sedimentary rocks of marine and non-marine origin overlie the volcanics.

The rocks, being pillow basalt, breccias and amygdaloidal basalts and andesites, were ready hosts to the mineralizing fluids that later deposited the various chalcedony gems. These volcanics are exposed throughout the west side of the state. For this reason gem hunting is very good in much of western Washington.

Lucas Creek has been a popular collecting area for many years. Large nodules of agate and jasper can be found in the gravels and banks of the creek during periods of low water. Rockhounds dig and screen the gravels and probe the stream banks or leisurely stroll along the gravel bars in search of the gems. Excellent agates have been found this way, but the best collecting is in the soil along the upper parts of the creek.

Digging is done in the rocky soil and alluvium in the forested areas of the upper 3 to 4 miles along the Lucas Creek Road. Some collectors spend most of their time probing the cuts of the numerous logging roads. Digging to a depth of about 3 feet yields most of the agates. The soil is mostly a sticky clay, sometimes rocky. A shovel and pick are the most useful tools; some rockhounds find that a screen can be used in some areas.

A great variety of material will be found, but most of it will be banded blue-gray agate nodules, carnelian nodules and agate and quartz geodes. Pieces of jasper of several colors, including browns, yellows, greens and reds, can be found.

Most of the land along the upper portions of the creek is owned by timber companies. Collecting is allowed on most of the land, but "no trespassing" signs should be obeyed. The valley along the creek belongs to several farms, and much of it is posted. Rockhounds can gain access to some of it by obtaining permission from the owner. Access and digging areas are shown on figure 8.

The North Fork of the Newaukum River to the north and west of Lucas Creek produces similar material. Gem quality material is not as abundant as along Lucas Creek, but some has been collected. Most of the material along the North Fork is located in the river gravels.

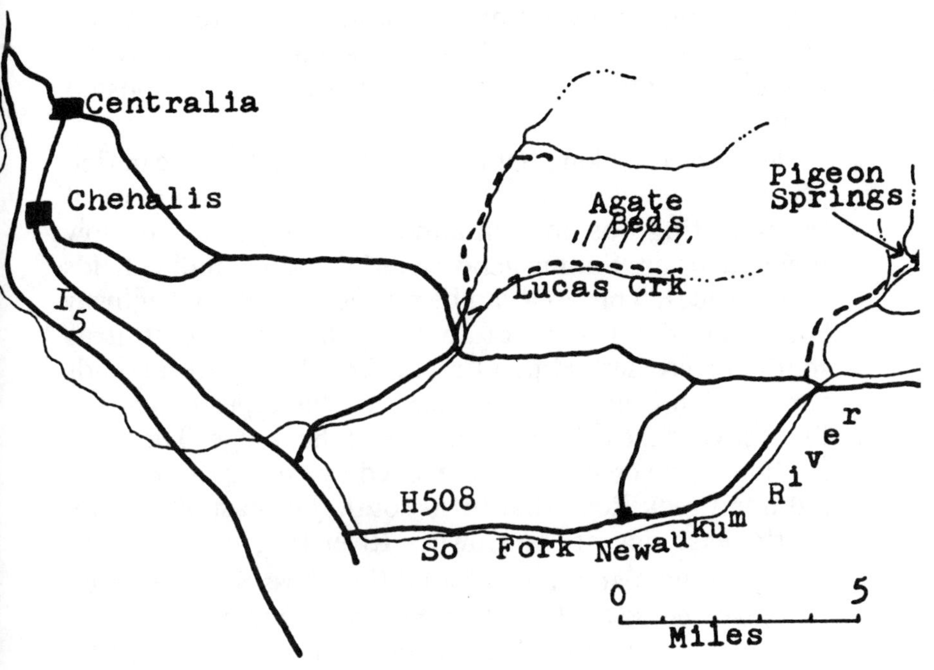

Fig. 8 *Lucas Creek-Newaukum River area*

The South Fork of the Newaukum River produces more material than the North Fork. Along the lower reaches of the stream, along Highway 508, most material is found in the river gravels. North of the Centralia-Alpha Road, agate and jasper are found in the river gravels, in the gravels of the tributaries and the stream banks. Frase, Bernier and Beaver Creeks are all good suppliers of material (Clevinger, 1970). Figure 8 shows the access to these creeks and to the Pigeon Springs locality at the end of the road.

Pigeon Springs is the most popular site in this area; it has produced some good quality carnelian and agate nodules. Some of the agate shows white plumes or moss. Jasper of several colors has also been found.

19

Collecting conditions are the same in this area as those in the Lucas Creek area. The land is mostly owned by the timber companies; collecting is allowed in the unposted areas.

Near Toledo, along Salmon Creek, the collector can find jasper, petrified wood and agates. In past years most of the collecting was done in and along the creek, but now rockhounds are finding good material in surrounding fields and hillsides. This area can be reached by taking Highway 505 east out of Toledo for about 6 miles to where it turns south and crosses Salmon Creek. From here, several roads can be found that provide access to the upper reaches of the creek, or the collector can park at the bridge. The lower portion of the creek can be reached by taking the Toledo-Salmon Creek Road that turns south, just east of Toledo, on the east side of the Cowlitz River (Fig. 9).

Numerous roads can be found that cross Salmon Creek and give access to the collecting area. During low water in the summer and fall good material can be found by searching the gravels and banks of the creek. Nodules of carnelian, agates and petrified wood, some of large size, can be found by carefully searching the gravels.

Along with the gem material most collectors will find that specimens of coprolites are abundant. The origin of these unusual specimens is widely contested. They may be fossil remains of animal dung. They are brown in color and ropy or nodular in shape (Fig. 10). Their composition is siderite, an iron carbonate. Although it is not useful as a gem, it is described here because the material is so abundant and of interest to the rockhound. The coprolites are probably eroding out of the glacial Vashon formation or out of nearby Miocene age sediments, both of which are cut by Salmon Creek.

South of Salmon Creek the rockhound will find good collecting in the Silver Lake area, which can be reached by following Highway 505 out of Toledo to Highway 504 and turning right to Four Corners. At Four Corners, Sightly Road is taken to the left (Fig. 9). This area can also

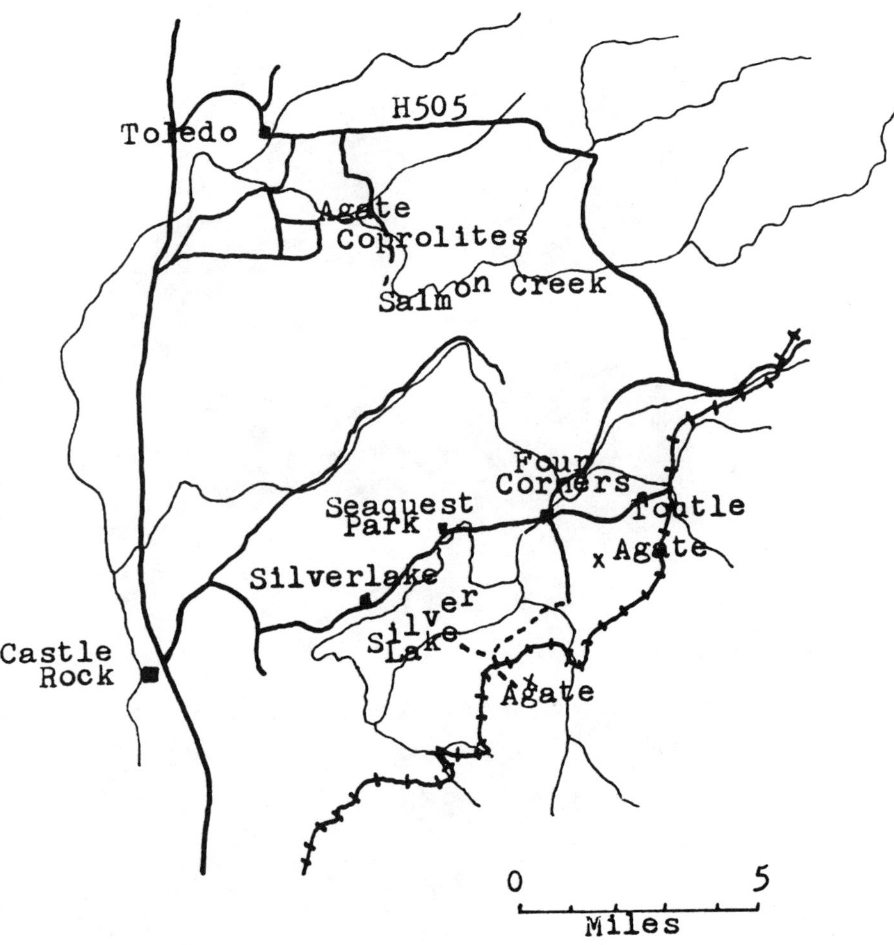

Fig. 9 *Salmon Creek-Silver Lake area*

be reached by taking Highway 504 east from Interstate 5 at the Castle Rock intersection, past Seaquest State Park to Four Corners. Some material may also be found near Toutle on the South Fork Road.

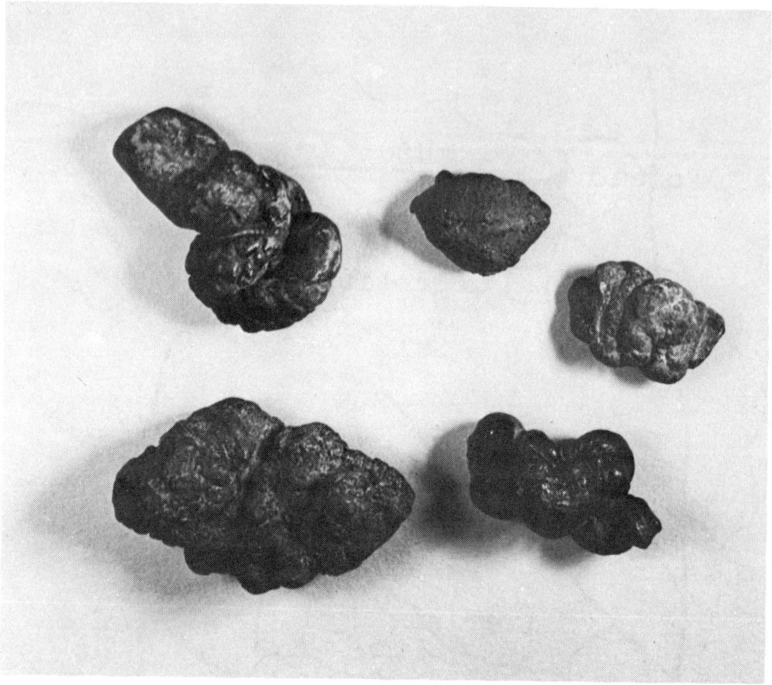

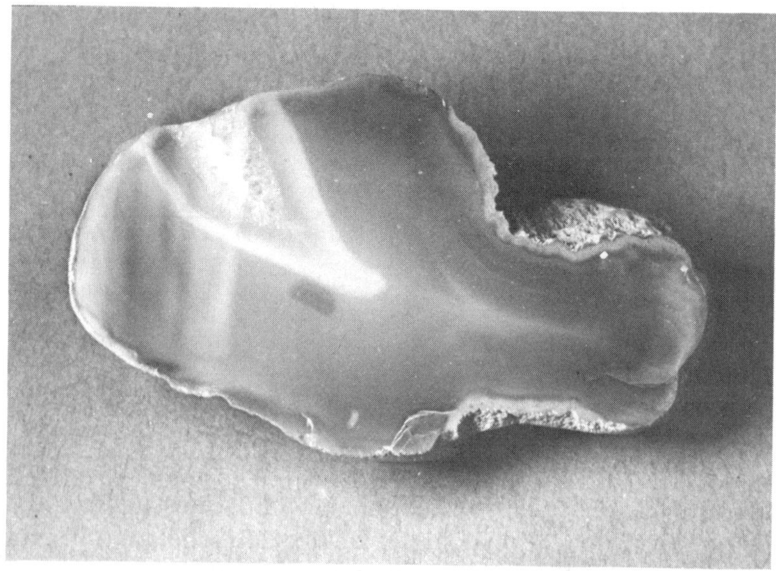

Fig. 10   *Coprolites, Salmon Creek*

Fig. 11   *Agate, Silver Lake. Al and Lucille Ockfen Collection.*

Agate, jasper and highly colored petrified wood occur in the rocky soil. Some specimens from this locality are illustrated in figures 11 and 12. The agate is often blue-gray or red in color. Jasper of several colors, along with petrified wood of a red, brown or yellow color may be found. The gems are often 4 pounds or more in weight.

Material is obtained by digging in the rocky soil and gravel pits. Most of it is found in the top 2 to 3 feet of the soil, but deeper digging will produce additional gems. Some collectors have had good luck searching along the railroad tracks. This area is quite popular and well searched, but still yields good material. The area at the end of the road is closed, but there are several road cuts, gravel pits and wooded areas that produce good gems.

About 10 miles east of Silver Lake on Highway 504 there is another good agate locality along Beaver Creek near Kidd Valley Park. The creek is a tributary of Green River that flows into the Toutle River at the park, which is an excellent place to camp. The collecting is on Line 516 on Weyerhauser property, access can be obtained by contacting the Weyerhauser Company in Centralia.

Digging conditions are similar to most of the previously described areas in this part of the state.

Three miles east of Interstate 5 on Highway 12 is the small town of Mary's Corner. Road cuts and gravel pits in this area contain some good pieces of carnelian and jasper. There are numerous places to search for material here.

The entire area from Silver Lake to Lucas Creek is underlain by the same basalts and related volcanics and sediments. It is probable that there are other collecting localities in this region that have not been discovered yet. There are miles of logging roads that are open to the public and provide thousands of road cuts for the ambitious rockhound to prospect in. More good material will likely be discovered in this area in the future.

Myrikite can be found in the mercury mines in the Morton area. This is good quality translucent to white chalcedony containing specks and streaks of cinnabar. It

Fig. 12   *Carnelian, Silver Lake. Al and Lucille Ockfen Collection.*

Fig. 13   *Banded Agate, Kalama. Al and Lucille Ockfen Collection.*

is associated with the ore minerals in the mines, in the volcanic rocks around the mines and on the dumps of the mines. It is scarce, but makes unusual specimens and can be cut into beautiful gems.

There is a large collecting area near the small town of Kalama on Interstate 5. The digging area is about 5 miles east of town on Green Mountain. It can be reached by taking the Green Mtn. Road out of Kalama to the top of the mountain (Fig. 13 and 14). Most of the digging has been done in the soil amongst the trees, but good material has been found in road cuts and in the gas line right of way.

Most material is found in the top 3 feet of soil. The soil is a clayey-loam; digging is not difficult. Most collectors find that a shovel is most useful, but a pick or mattock may be of use in many places. Some rockhounds screen the soil to find the small pieces of agate and jasper. This is not necessary, because most of the good material is of sufficient size that it is readily found in the soil.

Agates, quartz crystals and geodes are most abundant in this area. Some specimens from this locality are illustrated in figure 15. Most abundant is fortification and banded agate of blue and blue-gray colors. Some sagenite has been found (Fig. 16), but it is rare. Geodes lined with quartz crystals or with botryoidal chalcedony are less common than the agate. Loose crystals of quartz, probably from broken geodes, can also be found, 1 to 2 inches in legnth. Part of each crystal is clear, but fractures are usually abundant.

Most collectors will find that this is an excellent locality in which to collect. Agates cover a large area so there is ample room for everyone, for many years to come. A spectacular view of the Columbia River Valley can be seen from the mountain above Kalama. Digging is generally easy, even in wet weather. Most of this soil does not ball-up as so much of the clay in western Washington does.

Small enhydros of agate can be found on the edge of Kalama. The locality is in the basalt outcrops on the northeast edge of town. The small enhydros are quite abundant in this vesicular basalt. Most of them are small, being less

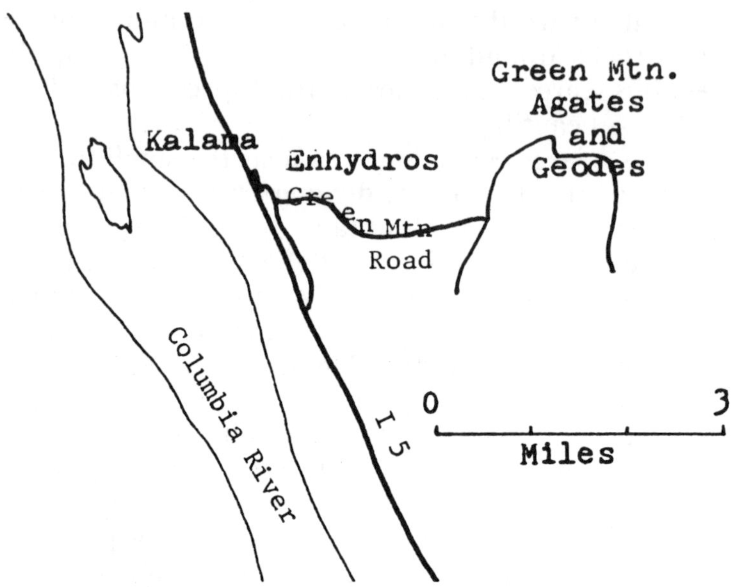

Fig. 14 *Kalama-Green Mountain area*

Fig. 15 *Banded agate and geode, Kalama.*

than .75 inch in diameter. They are a light gray to light yellowish-gray chalcedony that contain water, often with a small gas bubble. If they are tumbled, the bubble can be easily seen. Care must be taken when tumbling, as it is easy to remove too much of the agate, allowing the water to leak out. Only as much rough grinding as is necessary to smooth the enhydro should be done.

These interesting enhydros must be removed from solid rock. A hammer and gad are the most useful tools. By carefully breaking the basalt a collector can obtain several of these unusual specimens in a short time.

Similar enhydros are found at Spencer Creek, southeast of Kalama. They average about 1 inch in diameter, but a few are up to 4 inches. Approximately half of them contain water. A dark rind must be removed before the bubble and water can be seen. Care must be taken in this tumbling process as mentioned above for the enhydros at Kalama.

Collecting is done by carefully breaking basalt with hammers and chisels to remove the enhydros. This area is reached by taking the Kalama River road for 1.3 miles east out of Kalama to the Modrow Road. Turn right on this road and follow it for .5 mile and turn right on the Spencer Creek Road which is followed for 1.3 miles. The enhydros are found in the cut on the east side of the road at this point.

Old coal mines near the town of Tono have been good producers of gem material. Most abundant here are blue-gray fortification agates, jasper of many colors, petrified wood, carnelian, carbonized wood and rarely fossilized bones.

Some very good agates over 15 pounds in weight have been found in this area. Most of the agates are blue and blue-gray in color but carnelian is common. Some nodules are smooth and rounded, but others are very rough on the outside. Jasper is found in many colors including green, brown, yellow, red and black. Pieces are commonly over 3.5 pounds in weight. Most of the petrified wood that is

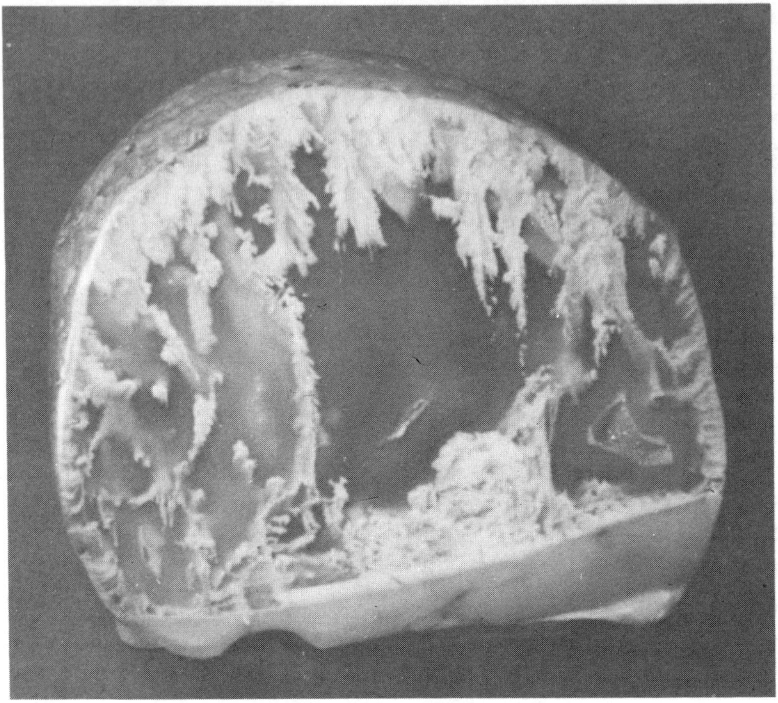

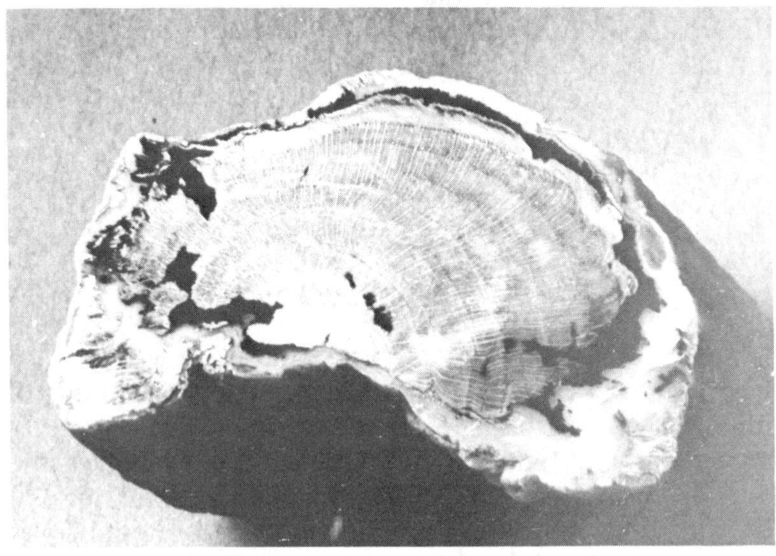

Fig. 16   *Sagenite agate, Kalama. Al and Lucille Ockfen Collection.*

Fig. 17   *Petrified wood, Tono. Al and Lucille Ockfen Collection.*

28

found here is not of gem quality, since it is usually poorly silicified, light brown or tan in color and full of small holes. The interior of the pieces is usually darker than the outside. Good quality pieces of wood replaced by brown to reddish-brown and black chalcedony can be found (Figs. 17, 18 and 19). Most of it shows the wood texture very well.

Since this is a coal producing area, carbonized wood is common. Some good specimens that retain the shape of the original tree can be found. The material is rather fragile, and cracks and disintegrates as it dries out. It does not make good specimens for display, but it is intersting to see. A few bones have been found with the coal. These are rare; the types and chararcteristics of the bones are not well reported.

The material is collected by digging in a brown clay. In some areas the clay is sandy and not difficult to dig in, but most of it is very sticky when wet. It balls up on shoes and shovels, and is difficult to wash off of specimens.

In some places where specimens and rocks are not abundant, rockhounds use steel probes to locate a likely specimen before digging for it. This area is very popular and has been widely dug. There are some deep holes full of water, so all collectors should use caution before stepping into a pool of water.

The main digging area is on the hillside on the northwest side of the valley (Fig. 20). It can be reached by taking Highway 507 out of Centralia and turning east on the Hanaford Valley Road. Proceed past the Centralia Coal Electric Plant and cross the valley. The old coal mine is on the left side of the road just after it changes from blacktop to gravel. Digging is done on both sides of the road.

Collecting here is done in the overburden from the mine and in the area around the mine. There are numerous holes in the woods above the mine and in the young alders that are growing on the original bulldozed area around the mine. One spot is as good as another. Some people have spent several hours digging here without finding very much of good quality. Other people have been very lucky

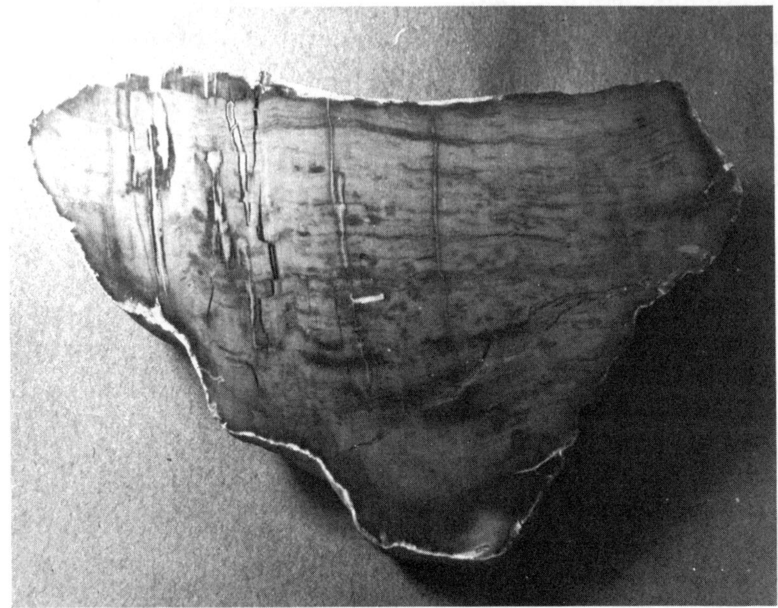

Fig. 18 *Petrified wood, Tono. Al and Lucille Ockfen Collection.*

Fig. 19 *Petrified wood, Tono. Al and Lucille Ockfen Collection.*

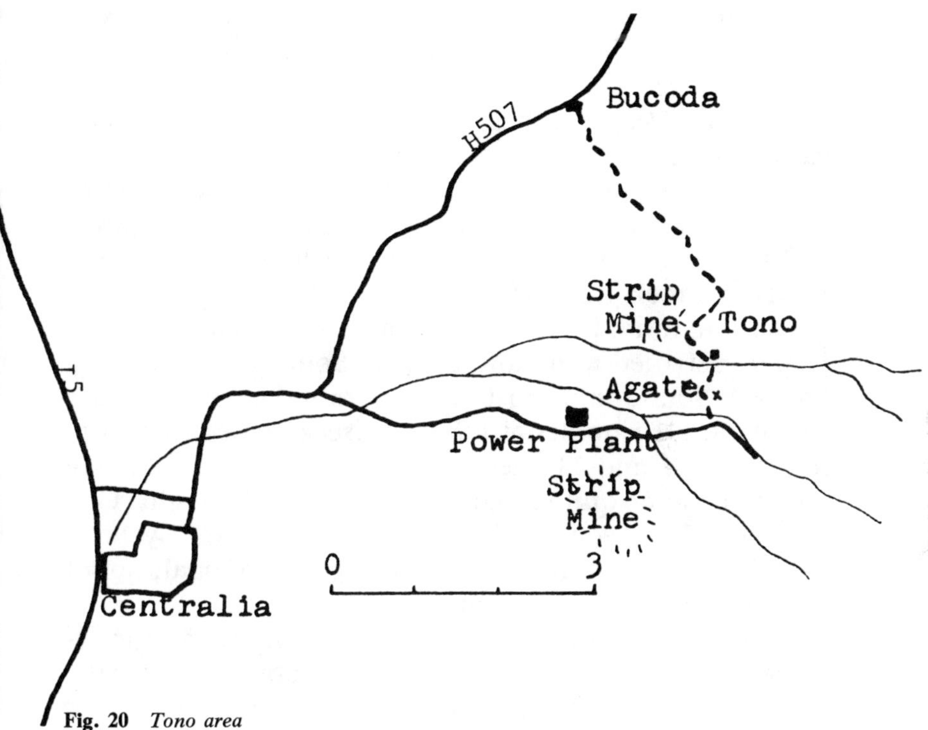

Fig. 20 *Tono area*

and hit a good pocket of agates most of the times that they try their luck at this site. In recent years it has become increasingly difficult to distinguish between ground that has and has not previously been dug. Even with this difficulty, Tono is still a good producer of gems.

On the other side of the road collecting is done in the soil around the creek. Digging conditions are about the same in both areas.

Miners for the Centralia Coal Plant are strip mining vast areas of the surrounding countryside. It is likely that some good agate beds are being exposed by the mining activities, but collecting is not allowed in the mining areas at this time. Because the miners are using large drag lines, bulldozers and trucks, there is much danger to the inquisitive rockhound. As each area is mined-out it is reclaimed. This may cover up any material that was exposed or it may leave some of it in the top few inches of the soil. Possibly, this area will be opened to collecting in the future.

Johnson Creek, southeast of Tenino, has produced good carnelian, jasper and geodes. This area is reached by taking the Johnson Creek Road off Highway 507 east of Tenino, or by taking the Skookumchuck Road off Highway 507 three miles south of Tenino and following it to the Johnson Creek Road (Fig. 21).

Gem material found here includes dark red carnelian, jasper, petrified wood and geodes. Some good carnelian has been found, but most of the dark red color is only on the surface. Examples of Johnson Creek agates are shown in Figures 22 and 23. Pieces of jasper of several colors are common, some quite large, over 4 pounds in weight. Colors include green, red, brown, mauve and yellow. Bloodstone has been reported, and pieces of opal, not of gem quality, have been found.

Most of the land is farm land, and permission is required to collect on it. Collecting is also done in road cuts, in plowed fields, along the gas line and under the power line. Material can be found on the surface in the road and power line cuts, but digging is required to find most of the good gems. The soil varies from clay to rocky loam, and most material is found at a depth of less than 4 feet. A shovel and pick will be useful. Collecting can be done year round in this locality, but digging may be difficult in some areas when the ground is wet.

Chalcedony, containing minutely banded green jasper, can be found in the canyon of the Skookumhuck River south of Vail. Search the stream gravels in this region for other chalcedony material. This area is above the reaches of the reservoir behind the Skookumchuck Dam. Beware of the beach around the reservoir where a thick layer of mud is forming. This forested land belongs to the Weyerhaeuser Company. Access is generally allowed through Vail if the gates across the roads are opened.

This is a large area that may be of interest to the ambitious rockhound with prospecting instincts. There are

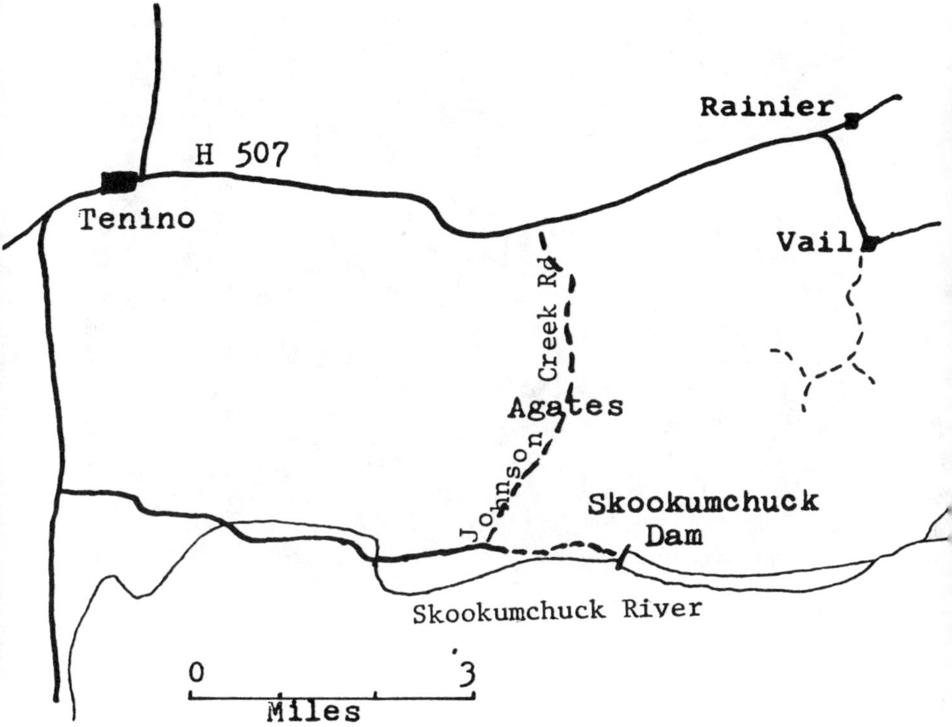

Fig. 21   *Johnson Creek area*

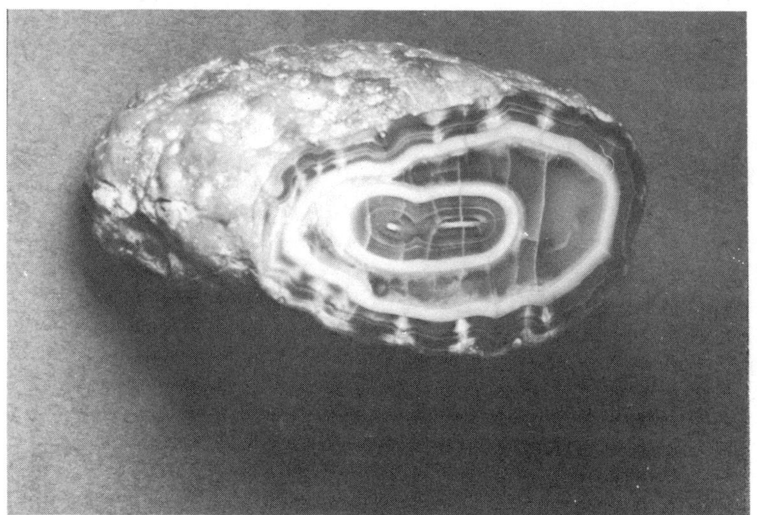

Fig. 22   *Banded agate, Johnson Creek. Al and Lucille Ockfen Collection.*

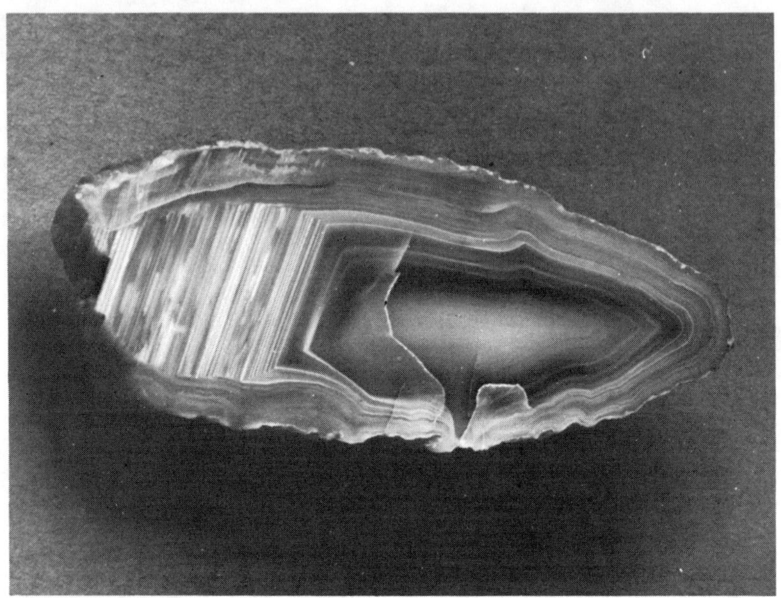

Fig. 23 *Banded agate, Johnson Creek. Al and Lucille Ockfen Collection.*

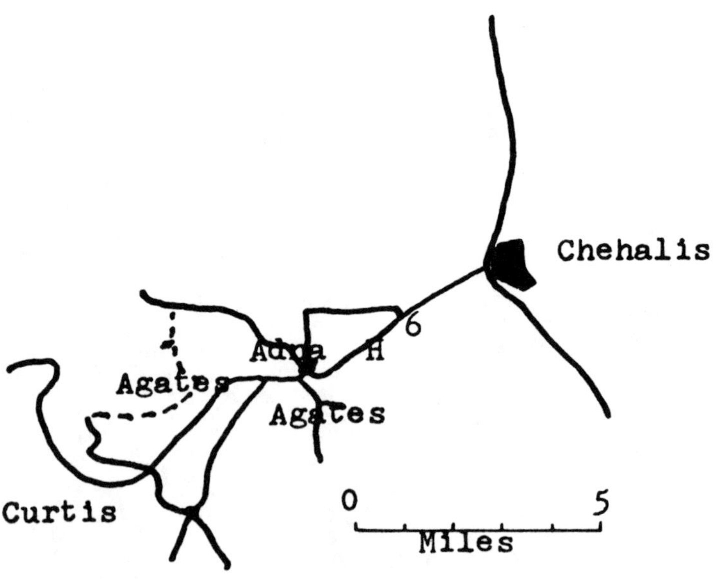

Fig. 24 *Adna area*

ample road cuts and quarries in the basalts to keep most prospectors busy for a long time. Various chalcedony gems may be discovered along with some zeolites for the crystal collector.

The Chehalis River has produced some good quality gem material. A rare piece of petrified wood, carnelian or agate may be found by rockhounds that search the gravels of this river. The best collecting is during the summer and late fall when the water level is low.

High quality material has been found in the Adna area west of Chehalis on Highway 6 (Fig. 24). This is a large farming area with several collecting localities. Material is found in the gravels of the streams and in the soil of the farms and forested hillsides. At various times, certain farms have been open to collecting. Some land owners charge fees, others do not. Collectors annually go to this area during the spring plowing season to take advantage of the exposed material in the newly-plowed ground. Rockhounds who are interested in collecting in the Adna area can obtain information and permission to collect from the local land owners.

Good collecting can be done along the road to Curtis that goes south from Adna. Material has been found in the road cuts and in the soil dug up by the construction of a water line in this area. Several large agates were found, including a 60 pound agate along this waterline (Fig. 25). The Adna pit (rock quarry) on the south side of town has produced some good material. The site is not always open so inquire locally for information on this good collecting locality.

Ceres Hill west of town, has produced some good material (Fig. 26). This area is reached by taking the road that goes west from town on the north side of the river for several miles to the Ceres Hill Road that turns south. This gravel road is followed for about 2 miles to a logged-off place. The collecting area is on the left side of the road just before the road makes a sharp turn to the right.

Digging is done within sight of the road. Most material is found in the top 3 feet of the clay soil, which contains

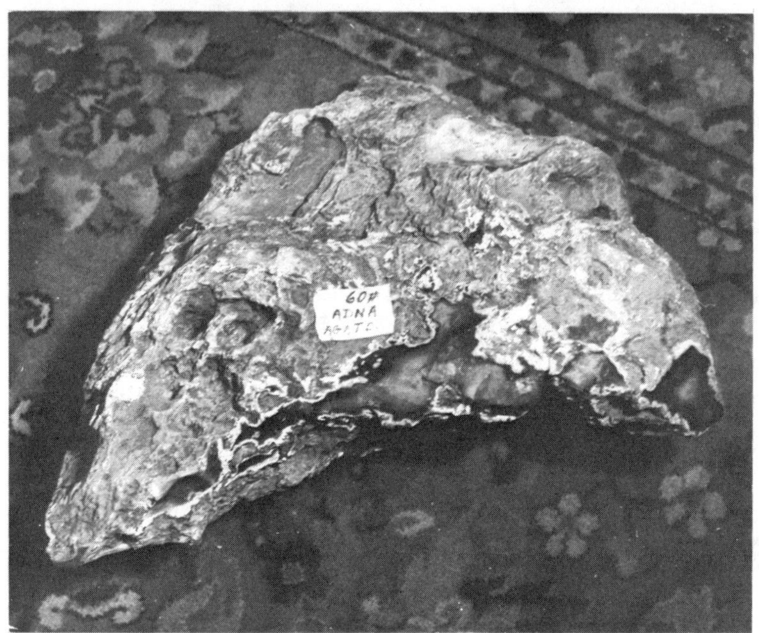

Fig. 25 *Unusually large agate, Adna. Al and Lucille Ockfen Collection.*

Fig. 26 *A lone collector at the Ceres Hill agate beds, Adna.*

abundant small pieces of agate and jasper along with larger agates. In wet weather, the clay can be very sticky and difficult to dig in. A shovel is needed, and a pick may be useful.

Some material is found in much of the surrounding logged area and in the road cuts. Very little digging has been done outside of the main area.

West of Adna on Highway 6 is the Green Creek area, a few miles west of Lebam (Fig. 27). A gravel road turns east off the highway and follows Green Creek for a few miles. Most of this land is private and has been open to collecting. Rockhounds have done some damage to the land, so be cautious and help keep this area open.

White to yellowish-gray, gray, bluish-gray and light red chalcedony casts of pelecypods, gastropods, nautilus and coral can be found in the gravels of the creek. Some of these are illustrated in figure 28. Fossils are not generally mentioned in this book, but the chalcedony material of these is very colorful and can be used as gem material. Whole fossils can be polished, the coral slabbed, or the pieces can be tumbled or cut into cabs.

The pelecypods are up to 2 inches in length, but average about 1 inch. Most of these are hollow and may be filled with water. Usually the water will leak out but in the rare specimen, it may remain. The gastropods are scarce but are found up to 1.25 inches in length. The nautilus, genus Aturia, casts may be more common than the gastropods; they certainly are larger. These unusual fossils range in size from less than 1 inch in diameter to over 3 inches in diameter. Each section of the nautilus may be a different color of chalcedony. The coral is a colonial type with white to buff septa and clear to white chalcedony filling the spaces between the septa. A few of the casts have some of the original shell remaining.

These specimens are found in the lower parts of the creek below the farm at the end of the road. To the author's knowledge there has been no intensive search of the surrounding country-side to find the source of these fossils. They are found in pockets in the gravels, concentrated near the bottom of the creek's gravels, much the same as gold is. The pockets are located by a combination of skill and luck.

They will be concentrated where the water makes a quick change in velocity, such as the inside of the sharp bends. Digging is required with a shovel and screen. Hip boots are needed as any hole dug in the creek rapidly fills with water, and the collector may be required to dig deeper than three feet. To find the material, a rockhound picks a likely spot and digs a hole. The gravel is screened and searched for the casts. If the material is not found upon reaching bedrock or a depth where the collector feels unsafe to go further, then the hole should be filled in and a new spot selected.

In most places only a few specimens will be found, but some bonanzas of over a thousand have been found in one hole. Because of the required digging in the creek, collecting is not possible during most of the year. A trip to this locality should not be planned before June or after September.

Good agate has been found in the Grays River near the town of Grays River on Highway 4. The agates are banded blue and blue-gray, a rare piece of carnelian has been reported. The material is scarce, but the fishing is good. The gems are found in the river gravels most of the length of the stream. It is accessible by roads along most of its length. During summer months gravel bars are large and abundant.

Large pieces of clear to black chalcedony have been found on Weyerhaeuser property near Enumclaw. This is a new find along road 1080C. The agates are found in large pieces, some of which are in excess of 20 pounds in weight. Small holes are numerous in parts of the seams of agate, but good material is abundant.

Northeast of Mt. Rainier on Weyerhaeuser property, there have been recent finds of good blue agate nodules and thundereggs. Very little collecting has been done in this area and access may be restricted because of logging activities. This area should be a good producer of excellent

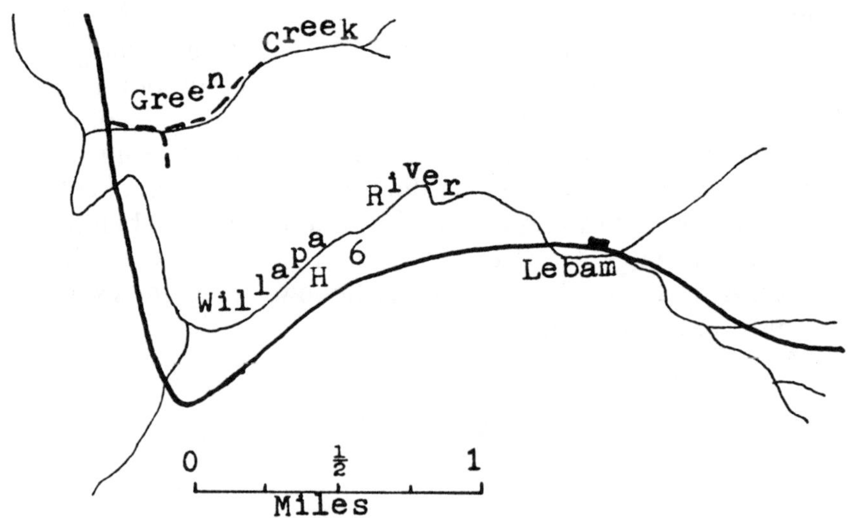

Fig. 27 *Green Creek area*

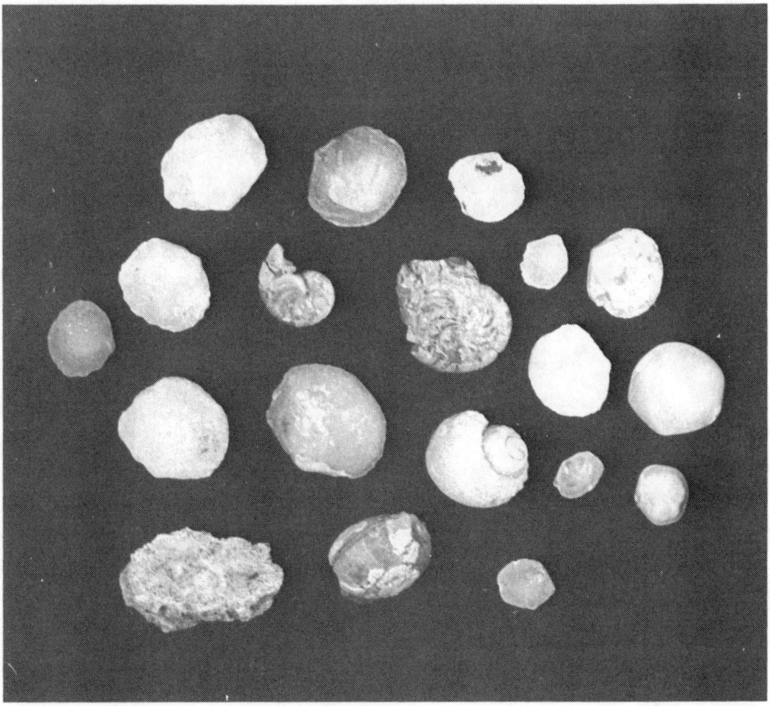

Fig. 28 *Chalcedony casts of nautilus, pelecepods, gastropods and coral, Green Creek.*

material.

The material is found in the soil weathered out of volcanic rocks. Most of the nodules average about 1.75 inches in diameter, and are good solid agate. The colors are of various shades of gray to blue and white. Some of the finely banded specimens show a good spectrum of colors when sliced thin and viewed with a bright light behind (Fig. 29). They are truly good iris agates. Most of the specimens are solid, but a few are hollow, being lined with small quartz crystals.

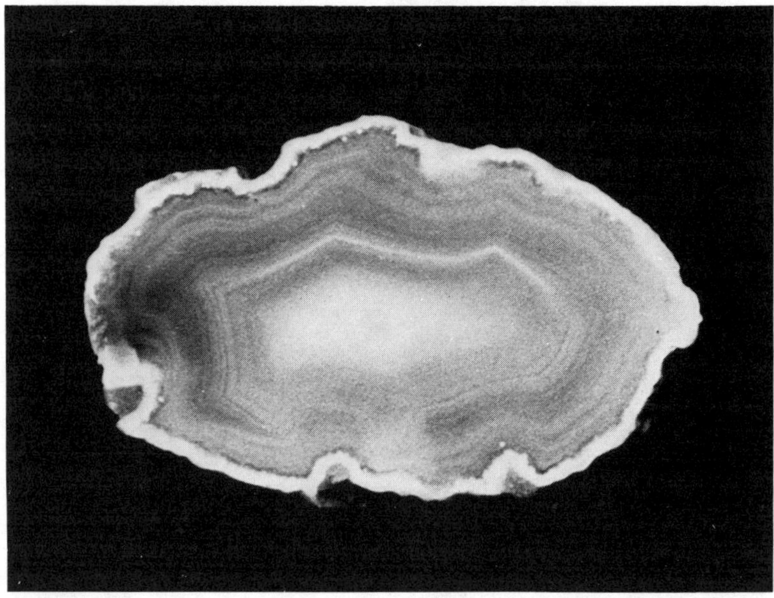

Fig. 29 *Banded agate, Mount Rainier.*

Northwestern Washington

Petrified wood has been found in Cherry Creek south of Monroe. To reach this locality take Highway 203 south out of Monroe 3.5 miles to a road that turns east. This road is followed up the hill for 1 mile to a "Y." Take the right road, follow it to the next "Y" and again take the right fork. Follow this to the gate and park there. From the gate it is a 1 mile hike down to the creek.

Most of the wood is black and much of it contains streaks of marcasite. Pieces over 10 pounds in weight can be found. They occur in the creek bed and in the banks of the creek. A rock pick and possibly a shovel will be needed to collect this material.

Blue vein agate has been found in the Barn Hill Quarry in Monroe. Most of this material will fluoresce under an ultraviolet light. Pieces up to 2 pounds can be collected with the aid of a rock hammer. The quarry is open to collecting on weekends.

The agate is of good quality and polishes well, but it is not a very dark color nor does it contain interesting patterns or bands. However, the fluorescence of the material makes it an interesting specimen material.

South of Sultan there is an occurrence of good quality banded jasper. The bands are dark maroon, yellow, gray and black. Most of the bands are nearly straight. Large pieces, several pounds in weight can be found. It takes a good polish and is useful for cutting cabochons. This material must be removed from its enclosing volcanic rock; a hammer and chisel or gad are the necessary tools.

The deposit is reached by taking 311th SE out of Sultan for .8 miles then turning right on the Ben Howard Road and following it for 3.6 miles to the Cedar Ponds turnoff to the right. Take the left fork and follow the road for .3 miles, take the left fork for .2 mile then take the right fork for .2 mile. Here take the right fork and follow it up the steep hill .3 mile to the diggings in the road cut. The jasper vein carries from 6 inches to 2 feet wide and has been worked back into the road cut for about 20 feet.

Blue agate and quartz geodes have been found in the Walker Valley area east of Mount Vernon. The Washington Mineral Council has a lease on part of the occurrence, and the Skagit Rock and Gem Club has an adjoining lease. This area is reached by taking Highway 9 south from the College Way intersection east of Mount Vernon to the Big Lake store. From here it is another 1.5 miles south to the Walker Valley Road which turns left. This paved road is followed for 2 miles, after which there is a dirt road that turns to the right. This is followed for 1.5 miles to the collecting area (Fig. 30).

The collecting area starts where the road forks, but the most popular area is along the right fork about .25 mile. There is a sign erected by the mineral council. The roads are county owned, so they are open to the public.

Agate is found in most of the exposed basalt. Some of the best material was found in the road after construction. The cliffs are exposed on the left side of the right fork. The first cut is a low basalt bank that contains small seams of blue agate. It is possible that the best material that was found in the road came from this spot.

Geodes over 12 inches in diameter have been removed from a basalt cliff. A good trail to the cliff begins at the mineral council sign. Amethyst is common, especially in the larger specimens. These geodes contain rhombohedral crystals of calcite and small blades of goethite on the quartz crystals. Because this material is best used as mineral specimens it is further described in the quartz section of part two of this book.

Caution must be taken when collecting here. If there is a collector at the top and another on the talus, then they must be careful that the lower party does not receive an unwelcome specimen from the collector above. Eye protection is necessary as small chips fly off of the basalt as it is being broken. Large hammers and chisels are needed to remove the geodes from this hard host rock.

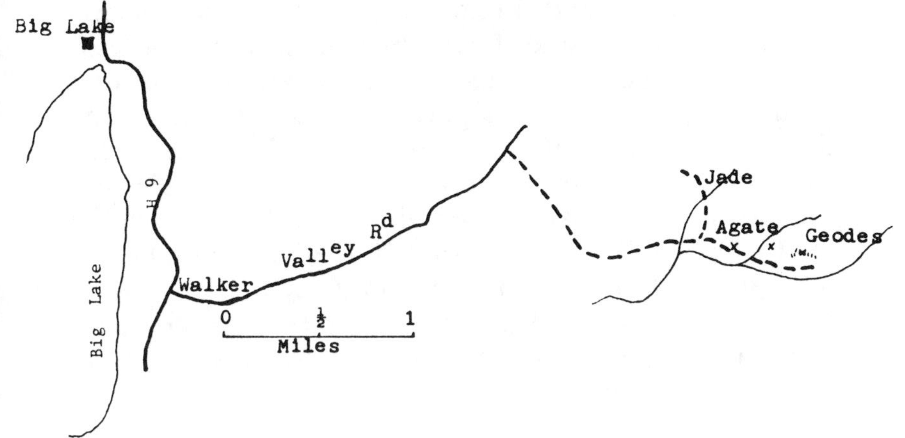

Fig. 30 *Walker Valley area*

Another blue agate vein is exposed in an old quarry to the north. At the point where the road forks, take an old road to the north that goes around a small ridge. Walk up the road to where it ends on the ridge. Take the trail to the quarry, walk up the draw to the left of the quarry, then follow the first draw to the right a short distance up to the next draw. The agate is found in the basalt on the left side of this draw. The color and characteristics are the same as those of the agate on the road. It does average a little better in quality.

The creek that is crossed by the left fork of the road is a producer of nephrite jade. There are boulders of jade in the creek that are more than 50 pounds in weight. Most of the jade is of a gray color, but some green does occur. Similar jade is found in much of the region between this collecting area and Darrington.

Other occurrences of blue agate have been reported in this general vicinity. There are numerous outcrops of volcanic rocks, any of which may contain agate or geodes. The locality described above is state land, but there is private land in the vicinity. Permission should be obtained before any propsecting is done on this land.

Agate nodules and thundereggs occur in the volcanic rocks on the north side of Mt. Higgins, northeast of Oso. Access is gained by turning north at the Whitehorse cross roads of Oso. Follow this paved road across the bridge to the gravel road that angles off and up a hill to the right, road no. 3403. Follow this road to Segelson Pass. Beyond the pass take the road to the right and follow it down the hill to the road that turns sharp left, road no. 3314. Follow it across Deer Creek then to Higgins Creek. Turn around and go back to the last big bend at the clear cut. The agates and thundereggs are exposed in the outcrops above the road at this point.

Olympic Peninsula-Puget Sound

Various agates and jasper can be found as pebbles on the beaches of the San Juan Islands and on the beaches of Whidbey Island. This material probably has been deposited by the glaciers that carved out the sound. Most of the pieces are small, less than a few ounces in weight, but larger specimens have been found. Collecting is done by walking the gravelly beaches during low tide. These beaches are most productive after a storm.

The material includes carnelian, banded agate, pertrified wood and red jasper. These are all well-rounded semi-polished pieces. Most abundant are small pebbles of jasper and agate.

West of Lake Crescent on the Olympic Peninsula, large pieces of spherulitic jasper have been found in pillow basalt. Most of it occurs in and around several manganese mines that can be found in the area. The jasper has been found in several areas on the peninsula. All but the north end of the Lake Crescent lies in the boundaries of Olympic National Park. No collecting is allowed in the park.

Some pieces of jasper weigh more than several tons. Outside of park boundaries, smaller pieces can be found in the gravels of the streams and the ocean beaches. Rialto Beach across the Soleduck River from Lapush has produced some of this material. Small pieces of the jasper can be found in the numerous gravel bars on the beach. The jasper is composed of red spherulites in a gray, green or translucent groundmass. Most of the jasper pebbles are less than 1 inch in diameter, but larger ones have been found (Fig. 31). At times this material is quite abundant, other times it is quite rare. Winter storms continually change the beach. Some summers the gravel is almost buried by sand, yet in other years there are high gravel banks on the shore.

Specimens consisting of bright red jasper spheres in a black manganese groundmass are found 7 miles west of Lake Crescent (Jackson, 1975). The Jasper weathers out of a brushy hillside about .5 mile north of the highway. The basalts in the area weather a red color also, so chip the pieces to see if they are jasper.

To get to the locality take Highwy 101 west from Lake Crescent. At 2 miles you pass the road to the Soleduck Hot Springs, follow the highway for 5.3 miles further. At this mileage point there is a stump with a red X painted on it between telephone poles 32-1 and 32-2. There is a parking area 300 feet west on the south side of the highway. Walk up the old road that goes north past the stump to the north side of the hill. Follow the base of the hill northeast to a small stream, then follow it north, always taking the left fork. The jasper is found in the rock outcrops on the east side of the stream.

South of Westport there has been good material found in the scarce gravels of the beach. The best beaches are at Westhaven Park. Again it is not always a good place to collect. Some storms bury the gravel with sand, others uncover the gravels along with a fresh supply of gemmy pebbles. Some examples of this material are illustrated in figure 32. Colorful agates, jaspers, and petrified wood are represented by these pebbles. Some of them are

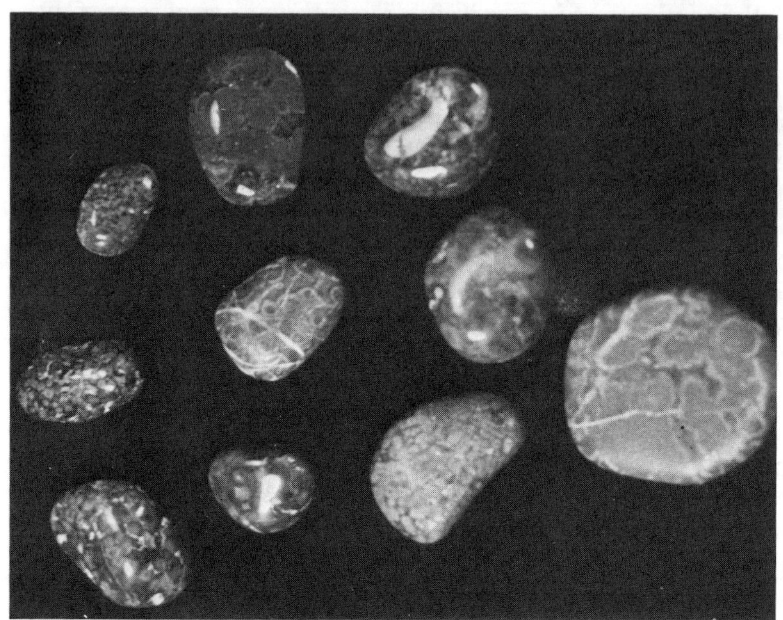

Fig. 31 *Spherulitic jasper, Rialto Beach. Largest specimen one inch across.*

Fig. 32 *Beach pebbles of agate, jasper, and petrified wood, Westport.*

46

suitable for cutting cabochons, others are best tumbled.

The gravel beaches on the west side of Puget Sound in Jefferson County have produced good material (Williams, 1974). Agate and jasper has been found on the beaches from Ludlow Bay to Mats Mats Bay, East Beach, and North Beach, of Morrowstone Island, and the beach northwest of Point Wilson at Port Townsend. These beaches are best found by following a state road map, a description here would be long and possibly confusing. The area can be reached by taking Highway 101 north from Shelton, or Highway 104 west over the Floating bridge from Edmonds.

Material found will include carnelian, petrified wood, jasper, and black jade. Masses of green grossular garnet have also been found. Collecting is done at low tide, no tools are needed. This area offers good collecting, camping, and sight-seeing for the entire family.

An abandoned quarry nearby offers some good collecting. The quarry is reached by taking the highway south from Port Ludlow to the intersection where the road to the left is taken. Follow this road to the first unpaved road to the right. At the end of this road is the quarry. In the loose volcanic rock of the quarry there is jasper and agate of several colors. Most collecting can be done with a rock hammer, but beware of overhangs and steep slides. All quarries should be treated as if they are dangerous.

There are abundant volcanic rocks on the east and north sides of the Olympic Peninsula. Much of the area is covered with logging roads and is open to the public. Part of it is private, and part of it is National Forest. There are few reports of collecting material in this region, but the possibilities are good. The ambitious prospectors and curious rockhounds may find this an interesting and rewarding area to prospect for chalcedony.

CORUNDUM

Corundum, with a hardness of 9, is the second hardest mineral known and has long been prized as a gem. Corundum can be found in many colors from black to white, and opaque to translucent. The opaque and barely translucent varieties are not very useful as gems, but some are cut into cabochons. The translucent stones are highly sought as gems; ruby is red corundum and may be the most highly prized of all gems. All other colors are sapphire, of these pink and blue are the most common, but yellow, orange and green are also known.

Corundum is not well known in Washington. The mineral occurs in pegmatites and metamorphic rocks, both of which can be found in this state. There are numerous places where the geology could be right for corundum to have formed, but actual occurrences are scarce.

Pink and blue crystals, some slightly gemmy, can be found with the thulite at Tunk Creek. This locality is described under thulite. Search the canyon walls, quartz veins in schist and boulders in the creek for small crystals of corundum. A hammer and chisel would be needed to recover any crystals.

Similar corundum crystals are reported from a locality near Pullman in the NW¼ of sec. 18, T13N, R45E. This is just south of Bald Butte which was previously described under quartz. The crystals are reported to be found in the soil of a wheat field. A search of the road cuts and creek along the property by several people including the author has failed to confirm this report. A more thorough search of the locality may locate some of these gems. The crystals would have originated in the pegmatites of Bald Butte.

GARNET

The garnet group of minerals is well known to most rockhounds. All of the six common garnets: pyrope,

almandine, spessartine, uvarovite, grossular and andradite, can be found in gem quality crystals. With a hardness of 6 to 7.5 and a wide variety of colors the garnets
are quite suitable for cabochons and faceted stones. A
garnet may rarely contain inclusions of other minerals that
produce asterism when the garnet is cut in a cabochon
shape.

Almandine garnets from Idaho are well known for this.
There are several garnet deposits in Washington, but these
are mostly of non-gem quality. A deposit of almandine
garnets with a star-producing ability has been discovered
on Sloan Creek in Snohomish County.

The garnets occur in a hard mica and quartz gneiss (Fig.
33) of pre-upper Jurrassic age. This gneiss is a local occurrence in biotite and quartz gneisses which are common
in the North Cascades. There are no other deposits
reported in Washington, but the possibility is high because
of the abundance of similar rocks.

Characteristics of these garnets are similar to those from
Emerald Creek in Idaho. They are dark purple in color
and of trapezohedral and dodecahedral shape. The best
crystals average .4 inch in diameter, but a few larger ones
have been found. Many of the garnets are sugary-grained
or contain mica inclusions. Only four-ray star cabs have
been cut from these garnets (Fig. 34), but it is possible
that some of the crystals may produce six-ray stars. As
in the Idaho star garnets most of the star producing "silk"
is concentrated near the center of the crystal.

Crystals that are broken expose this "silky" portion of
the crystal, but most of the crystals at this locality are
whole. They are collected by digging and screening the
clay and gravel of several small creeks and gullies on the
mountainside. All but one of these small streams dry up
before the end of the summer.

The deposit is located in a scenic valley in the North
Cascades. During late fall, winter, spring and early summer the area is covered with snow. This locality has been
preserved for collectors by the Washington State Mineral

Fig. 33 *Almandine star garnets in gneiss, Sloan Creek.*

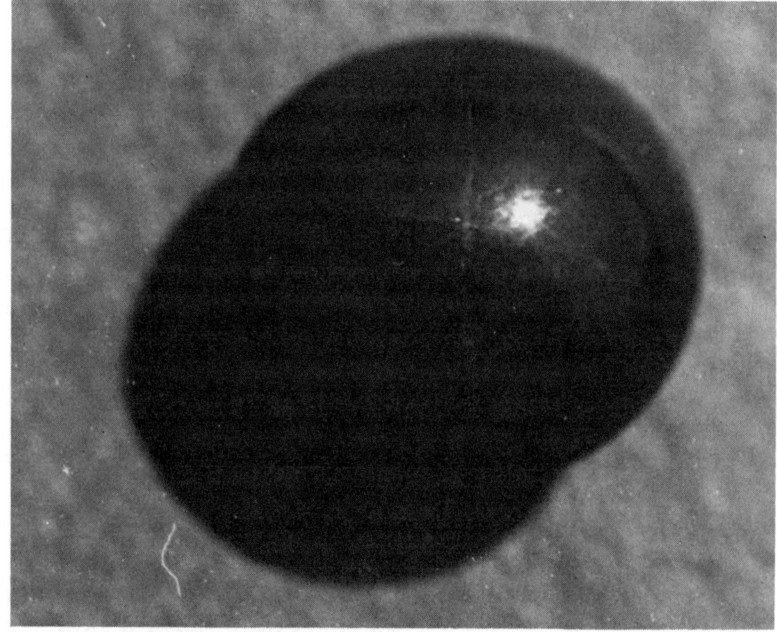

Fig. 34 *Star garnet cabochon, Sloan Creek.*

Council.

Massive gem quality grossular has been found with nephrite in Skagit and Snohomish Counties. Numerous pieces of this gem material have been found in Deer Creek north of Oso. It is found as water-worn pebbles and boulders up to and over 10 pounds in weight in the gravels in the creek. Nephrite is found in these same gravels. The garnet is white and white with green spots or streaks. It is semi-translucent and cuts and polishes very well.

Small pieces of the same material can be found in a branch of the Brooks Creek Road northeast of Oso. Most of these pieces of garnet are of small size, less than 1 pound in weight. They are found in the gravel of the lower 2 miles of the road. The road is on state land and has a locked gate at its junction with the Brooks Creek Road. Vehicles are not allowed through the gate, but some material can be found by walking along the lower part of the road. Beware of logging trucks and do not block the gate. Small pieces of nephrite can be found with the garnet. Most of this jade has a reddish-brown rind and yellowish-gray interior. It is useful in carving or cutting some jade cabochons of unusual color.

On the upper portion of Brooks Creek there is a vein of this garnet in situ in a chloritized serpentinite. The deposit is held under a mining claim. The vein is over 15 feet in width. Its shape is not determined; it actually may be one of several pods or lenses of the garnet in the area. Some production of this material has been done, but the remoteness of the area makes it difficult.

The grossular varies from white, to white with green spots and streaks. A few pieces of green translucent serpentine have been found with the garnet. Water-worn pieces of the garnet can be found in the creek below the vein, but the mineral rights are held under a mining lease from the state. No collecting is allowed.

HEMATITE

The iron oxide, hematite, is often used as a gem. Hematite crystallizes in the hexagonal system, but crystals are not common. It is more common as masses of either red or black color: red, if the masses are formed from very fine grained material, and black, if of large grain size.

Small dull, black, usually somewhat flattened masses, mostly under 3 pounds in weight, can be found in Finney Creek (Fig. 35). Red streaks are abundant in some pieces. Some magnetite grains are included, giving the specimens a weak magnetism. This characteristic is useful to distinguish the hematite fom the black graphitic schist that is found with the hematite and is much more abundant.

The best collecting area is reached by taking the Sauk River Road from Concrete. Cross the Skagit River and turn left, follow this road for 10 miles to the Finney Creek Road on the right. The collecting area is just after the first bridge on the Finney Creek Road, about 3 miles. Some good gemmy serpentine and rare nephrite slicks can be found in the creek about 6 miles further up the road.

JADE

Jade is a very popular gem that is rare as compared to most of the gems that are collected by the amateur. There are two minerals that are known as jade. Nephrite, a member of the amphibole group, is the most common. It is not a distinct mineral, but is a compact form of the actinolite-tremolite series with a calcium iron magnesium silicate composition. The common form of nephrite is irregular masses, usually rounded, but it also occurs in veins and as botryoidal masses. The masses are composed of minute fibers that are intergrown so that it has an extreme toughness.

Many colors of nephrite have been found, but green and gray are the most common. Specimens are opaque or

Fig. 35 *Collectors at Finney Creek, collecting nephrite and hematite.*

translucent. The slight translucency and its color patterns makes cabochons the best cut for nephrite. It is also widely used for carvings, where the extreme toughness gives strength to thin edges on delicate carvings.

Jadeite is the less common of the two jade minerals; a sodium aluminum silicate, it is a member of the pyroxene group of minerals. A few small monoclinic crystals have been found, but most jadeite is found as rounded masses composed of interlocking grains. Like nephrite, jadeite is dominantly green in color, but it is found in many other colors including blue, red, and white. The black variety contains iron and is called chloromelanite.

High heat and pressure in a metamorphic environment are required to form jade. It is usually found associated with serpentine, most often along contacts of a serpentinized ultramafic intrusion with gneisses, schists or marble. Geologically the most likely places to find jade in Washington are along the peridotite belt in the Central and North Cascades. This extensive belt of serpentine and peridotites may contain many yet undiscovered deposits of nephrite. Those deposits that have been found lie in this belt, except for the rare occurence that has been reported in northeastern Washington.

Prospecting for jade should be most productive along this belt and along the small serpentine bodies that lie outside of the belt. The largest serpentine and ultramafic bodies are shown in figure 36, but there are smaller bodies throughout this part of the state. Jade would most likely be found by prospecting the boulders and rocks of the creeks and rivers that flow off the western flanks of the Cascades from Highway 2 north to the Canadian border.

Most of the jade found in Washington has a weak-to-well-developed rind. Most of the rind looks slick and jade-like, but some of it does not. The rind may be gray, blue-gray, green, brown, reddish-brown or white in color. A hardness test in the field is usually not a good determination, because the rind of even the best jade is softer than the hardness of 6 for jade. Some of the jade is mixed with serpentine in the same boulder, and some of the boulders that look like layered, talcy serpentine on the outside have good jade on the inside. Practice and experience are necessary to determine if a boulder may be good jade, but it is difficult for even the seasoned collector.

Most of the deposits in the state, including the first few that are described here, are legally claimed for production of the jade. It is illegal to remove jade from these deposits without the owner's permission.

Nephrite has been found as float in several of the creeks and in situ in the area around Darrington. One of these deposits is located 8 miles west of Darrington. The Washington Gem Jade and Mining Company operates the mine. Collecting is allowed on a fee basis during the summer.

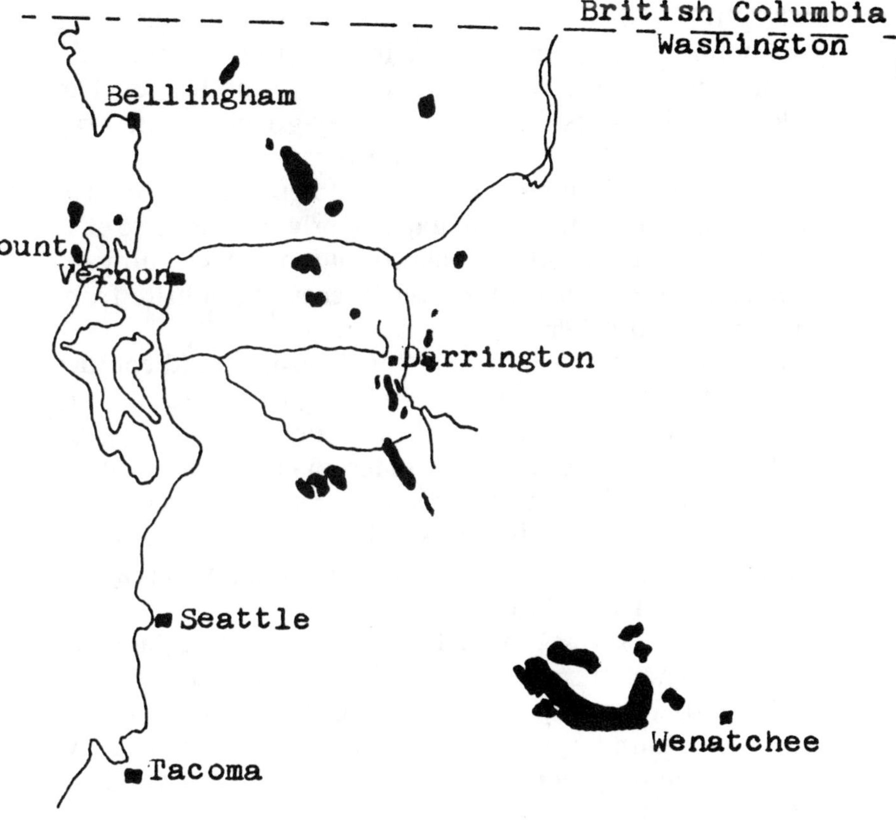

Fig. 36 *Map of major serpentine and ultramafic intrustions along the "Periodotite Belt".*

The deposit is in a serpentinite body in a steep ravine (Fig. 37). The host rock is a gray to green serpentinite, containing talc and chlorite. In some parts of the deposit there are long course fibers of tremolite. Nephrite is found as angular to rounded nodules up to and over 200 pounds in weight. Most of the jade pieces are from 2 to 20 pounds in weight. Figure 38 is of a large, about 40 pounds, nodule of the jade just as it can be found in the alluvium in the deposit with other pieces of nephrite and serpentinite.

Nephrite is so abundant in this deposit that it gives the rockhound no challenge to find it. Fractures are common in the small pieces, but solid pieces can be found. The color of jade varies from gray to gray-green to blue-gray to blue-green. Good pieces of translucent nephrite are not scarce. Pieces of blue-gray chatoyant nephrite are not uncommon. The fibers are usually poorly oriented, so it is difficult to cut good cabochons out of this chatoyant material. Figure 39 is of various pieces of nephrite from this mine and other deposits in the area.

Collectors who wish to visit this deposit should contact the Washington Gem Jade and Mining Company at Darrington. The deposit is not always open to collecting by rockhounds, but it is a good place to collect when it is. The collector will need a rock hammer and a shovel. Some jade can be found in the creek below the mine. There are numerous places to camp around Darrington; this is a very scenic part of the state.

Several miles west of this deposit the same company has claims on another serpentinite body that produces nephrite and grossular garnet. A few pieces of botryoidal nephrite have been found here too. Most of the nephrite is gray, but some good green has been found. The grossular is green to white, and white with green spots.

Most of the grossular is found in a vein in chloritized serpentinite, and as boulders in the creek. The nephrite is also found as boulders in the creek. No nephrite has been found in situ at this locality. It is not accessible by road, so no production has been planned from this deposit.

Several creeks near Darrington have produced a few jade boulders. The author visited one creek that has abundant boulders, many of them over 2 tons in weight. Most of the jade is gray to blue-gray in color, but good green is found with it.

Most of the jade has a slick, jade-like rind. Some of it does not look like jade on the outside, but proved to be good quality nephrite when cut by the owners of the claim. A small production of jade has been made from this creek.

Fig. 37 *Poor Boy Jade Mine. Jade nodules are found in aluvium in the ravine and in serpentinite at the top of the ravine.*

Fig. 38 *Jade nodules with serpentinite, Poor Boy Mine.*

A search for the source of this alluvial jade is very difficult. The creek is located in a narrow canyon that has a dense growth of underbrush, typical of western Washington. The source may be nearby or several miles up stream. The concentration of the jade near the lower part of the creek suggests that the source is not far up the canyon.

Another deposit nearby contains a vein of green and white patterned nephrite. Most of it contains small green flecks, probably grossular garnet inclusions. It is an unusual type of jade, but does not have a high degree of translucency, although pieces are fracture free. The pattern is unique for Washington. There has been limited production from this vein.

There are several small serpentinite bodies along the Skagit River. Jade has been reported from a few of the small creeks that feed the river near the serpentinites. Day Creek is reported to have produced several good pieces of nephrite. Likewise Finney, Presentin, Mill, Jordan and Boulder Creeks have produced some jade.

One of the creeks most popular with jade collectors is Deer Creek, a few miles north of the small town of Oso (Fig. 40). Large slicks of nephrite and grossular have been found in this creek. During periods of low water, July to October, large gravel bars are exposed. Jade of several shades of green, blue-green and gray can be found on these gravel bars. There are many miles of the bars to collect along and numerous areas to camp on this creek.

A few of the creeks that drain the Cultus Mountains, east of Mount Vernon, have produced some jade. Figure 47 is of a boulder of nephrite that was found in one of these.

Botryoidal nephrite has been found by the author in four localities. A vein of serpentine in a boulder of nephrite in a creek south of Darrington contains a layer of botryoidal nephrite. This is not the usual type of this material. The individual "bubbles" of nephrite can be separated from each other. The jade is good translucent green on the outside and white on the inside.

Numerous pieces of this rare form of nephrite have been found in a creek previously described as being held by mining claims. This botyoidal nephrite varies from gray-green to light gray. The individual masses vary from a few ounces up to about three pounds in weight.

The author and friends have discovered small occurrences of botryoidal nephrite in the Cultus Mountains, east of Mount Vernon. These deposits have varied in quality, and most of them have been low quality. Some of them produce specimens with individual "bubbles" of large size and some specimens are over 3 feet across (Figs. 42 and 43).

Other deposits of nephrite boulders have been found in this area. The unusual boulders and small slicks with reddish-brown, yellowish-brown or other color rinds are useful in carving. These slicks are much more common than slicks of high quality green nephrite.

Small pieces of alluvial jade can be found on the islands west of Mount Vernon. Black nephrite can be found on the many beaches of Whidbey and possibly Fidalgo Islands. The jade occurs with small pebbles of agate and various metamorphic rocks. Pieces of black serpentine and hornfels are abundant, making the detection of jade difficult.

The jade probably originated somewhere to the north and was deposited on the islands by continental glaciation. There are bodies of serpentinite on Fidalgo and Cypress Islands. There is a possibility jade may be found at these localities.

Similar jade has been reported on the beaches of Hood Canal south of Port Townsend. These jade pebbles have the same origin as those on Whidbey Island. These beaches are excellent places to spend an afternoon searching for gem material. The weather is often better here than in most of western Washington, because of the rain shadow of the Olympic Mountains. The scenery is excellent; on clear days the Olympic and Cascade Mountains can be seen in all their rugged splendor.

Fig. 39 *Various jades and pseudo-jades. Two pieces at top and piece at lower right are dark green nephrite from the Poor Boy Mine. Lower left specimen is green and white striped nephrite from Helena Ridge. Piece at bottom center is honey brown anthophyllite from Jumbo Mountain.*

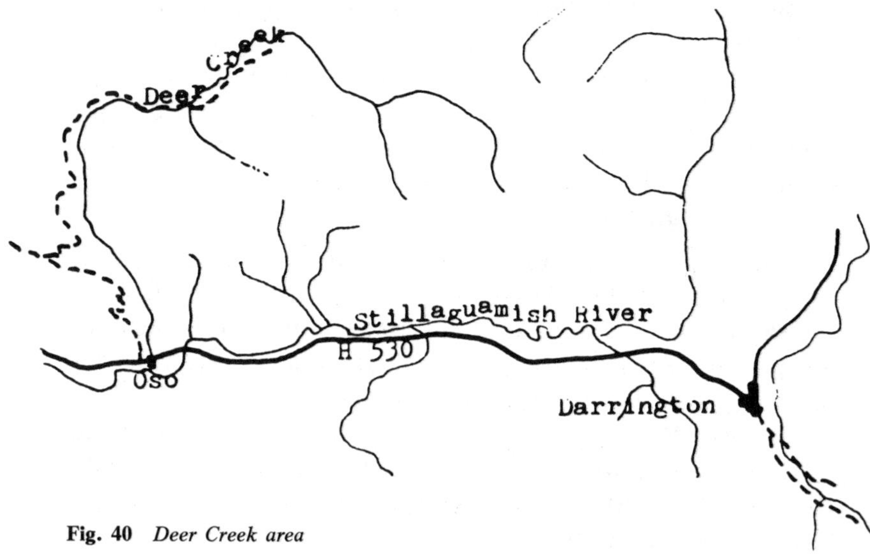

Fig. 40 *Deer Creek area*

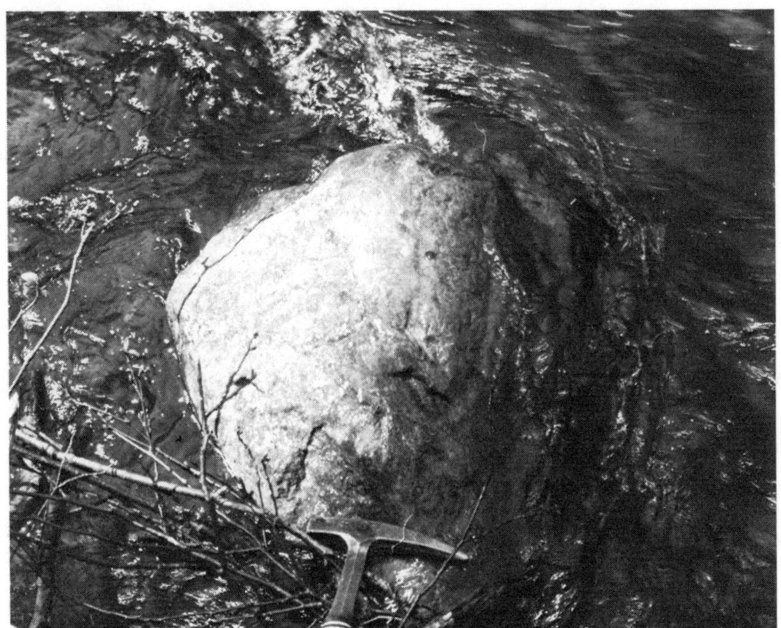

Fig. 41 *Light green jade slick in creek, near Oso.*

Fig. 42 *Botryoidal nephrite exposed in a serpentine outcrop. Cultus Mountains.*

61

Fig. 43 *Botryoidal nephrite, Cultus Mountains. Largest specimen is 3.25 inches.*

At least one deposit of nephrite has been discovered at the south end of the serpentinite belt at Mt. Stuart, which is located 20 miles north of Cle Elum. Nephrite from this locality is a mixed pattern of white, green, gray, brown and black. This nephrite is being mined and is available in limited quantity from dealers.

There are rare serpentinite bodies in eastern Washington, especially north of Okanogan. A few unverified finds of nephrite from these serpentines have been reported. The possibility of jade being found in this area is good. Collectors should do a little prospecting, as there may be a good deposit waiting to be discovered in this scenic part of the state.

There may be other deposits of nephrite and possibly jadeite in the serpentinite belt to the south and north of Darrington. Presently there is only a limited production of jade from Washington, but the state has a potential to be a large producer. Those collectors who enjoy prospecting for gem material will find that there are numerous streams along which to prospect in western Washington.

OBSIDIAN AND BASALTIC GLASS

Early man used obsidian to make projectile points; today it is a familiar material to most people. Its brittleness made it ideal for flaking into the desired shape for a point, but this same characteristic makes it a difficult material to work by the lapidary. Carvings can be made from it, however, extreme care must be taken to prevent breakage. Cabochons can also be cut from some of the various types of obsidian.

Generally obsidian is a black or brown and black volcanic glass. There are a few types of obsidian that have a silver, gold or rainbow sheen. These are rare and highly prized as gem material.

Obsidian forms by the rapid cooling of a lava flow that has a high silica content. The rapid cooling does not allow any minerals to crystallize, so a natural glass is formed. Obsidian cliff in Yellowstone National Park is a large and excellent example of such a flow.

Young volcanic rocks are abundant in the Cascade Mountains providing the proper environment for obsidian. In spite of this, few obsidian deposits are known in this state, and none of them are reported to be gem producers. Small pieces, mostly under 1 inch in diameter can be found, but rarely anything larger. The best places to look for obsidian are around the young volcanoes; Mt. Rainier, Mt. St. Helens, Mt. Adams, Glacier Peak and Mt. Baker.

Some basalt flows contain small quantities of a rare glass, basaltic glass. This material has little data recorded on it as a gem. It forms with palagonite when molten lava flows into a body of water and is rapidly cooled. Like obsidian the glass naturally devitrifies to clay, so only young flows would contain any glass. The pieces are mostly small, less than 2 inches long. Much of it is porous, but good solid pieces can be found. The color is dominantly black, but gray and blue glass does occur.

One location in southeast Washington has produced some excellent material for cutting cabochons. It is located in pillow basalt and palagonite about seven miles downstream from Lewiston, Idaho on the north side of the Snake River. This is about 2 miles upstream from the Moses precious opal deposit. Figure 4 shows the location of the deposit. The glass is found in the nearly flat-lying basalt flows near the tilted flows of the Lewiston monocline (Fig. 44).

Fig. 44 *Lewiston Monocline, folded basalt flows near the basaltic glass locality.*

The deposit is above the road on private property. Permission to collect has been allowed in the past as long as general courtesies are followed.

Park at the gate where the road crosses the railroad for the last time before reaching the corner at the precious opal deposit (you may desire to drive to the corner to be sure you have found this crossing). A hike of about 500 feet straight up the sloping canyon side is required to get to the diggings, which are barely visble from the road. As the collector hikes up the canyon the deposit becomes visible. An exposure of pillow basalt has been worked for about 30 feet along the side of the canyon. Digging is dangerous, all collectors should wear a hard hat for protection. The matrix is a palagonite and pillow basalt mixture, now partially altered to yellow clay. A heavy hammer and chisel or bar are the needed tools. The glass will be found as angular and rounded fragments less than 2 inches in length. Most of the nodules are coated with clay and an opal-like material of a light tan color.

This basaltic glass is black, brown, blue, lavender and white in color, and several colors may be mixed in one specimen. The colors occur in layers and swirls or solid in the pieces. The lighter colored pieces may be somewhat porous, but most of it is extremely hard and takes an excellent polish. Much of the glass is magnetic, especially the black, because of a high content of magnetite. Nodules of various sizes and shapes are scattered through this deposit.

Some calcite occurs with the basaltic glass, and there are some zeolites in the basalts below the glass deposit. The basalts of this area contain abundant chalcedony in many colors. Anyone searching for the glass should search the canyon walls for pieces of chalcedony and other vesicle fillings.

Other occurrences of palagonite in the state may contain good quality basaltic glass. Most of the petrified wood occurs in palagonite. When collecting agates and petrified wood in eastern Washington look carefully at those hard black nodules that you have been throwing away for years. They are not all basalt.

OPAL

Gem quality opal is abundant in the basalts of eastern Washington. Opal is an amorphous variety of silicon dioxide, with up to 10% water. This feature often causes cracking and crazing of some opals as it loses the water before or after removal from its matrix. Being amorphous, opal does not display any crystal shape unless it is pseudomorphous after another mineral. Usually opal is found as massive, reniform and stalactitic growths or replacing wood. Its hardness is only 5.5 to 6.5 and it is quite brittle. The prominent conchoidal fracture that develops when it is chipped made it a very useful material in the making of projectile points by early man.

There are several varieties of opal, many of which are found in this state. Common opal is the name that is applied to opal that is opaque or slightly translucent and of any color. It usually occurs in massive form; in water-laid basalt, it is often somewhat translucent. Hyalite opal is a water-clear variety found as stalactitic coatings in volcanic rocks. One characteristic of hyalite is that it often fluoresces when viewed under an ultraviolet light.

Precious opal and fire opal are the two varieties of opal that are the most used as gems. Any opal that shows a play of colors is known as precious opal. Some types of this are very high quality and thus very valuable. The translucent red to orange opal is known as fire opal, but this name is often applied incorrectly to precious opal.

Fire opal and precious opal best display their colors when cut as cabochons, but some faceted gems have been cut from these two types of opal. The common varieties of opal are often cut into cabochons or cut and polished as slabs. The color and patterns of opalized wood are best displayed when the wood is cut into slabs. Opal is useful in carving, but the brittleness of the stone makes it a difficult material to carve.

Opalized Wood

The most abundant gem quality opal in the state is the pertrified wood in central and eastern Washington. The Saddle Mountains are perhaps the best producers of this material. In this region of sage brush-covered basalt flows, opalized woods of many colors can be collected. Several kinds of trees have been identified from their petrified remains. These include redwood, oak, cypress, elm, maple, gingko and others. The wood is believed to be driftwood that was lying on the shores of large inland lakes. As the lakes were filled with advancing lava flows, the wood was accumulated in the forming pillow basalts and palagonite. Silica bearing waters seeping through the basalts replaced the wood tissue with opal, or in some areas, chalcedony.

The Saddle Mountains, southeast of Vantage, offer some of the best collecting. Much of the land is public land, so it is open to collecting. A few roads give access to the Saddle Mountains both east and west of the Columbia River. On the east side, wood can be found near the small towns of Corfu and Smyrna. Inquire locally for access and permission to collect on private land on the sides of the mountains. Local clubs can supply information on the exact locations of good collecting in this area. Figure 45 is a general map of the area; there are numerous side roads that give access to good collecting areas not shown on this map.

East of Mattawa, in the Saddle Mountains, some of the best opalized wood in the state has been recovered. This is the type that is known as Saddle Mountain Picture Wood. This locality has been preserved for the rockhound by the Washington Mineral Council.

Required tools include a bar or pick to loosen the basalt and a large hammer and chisels. Some collectors use a

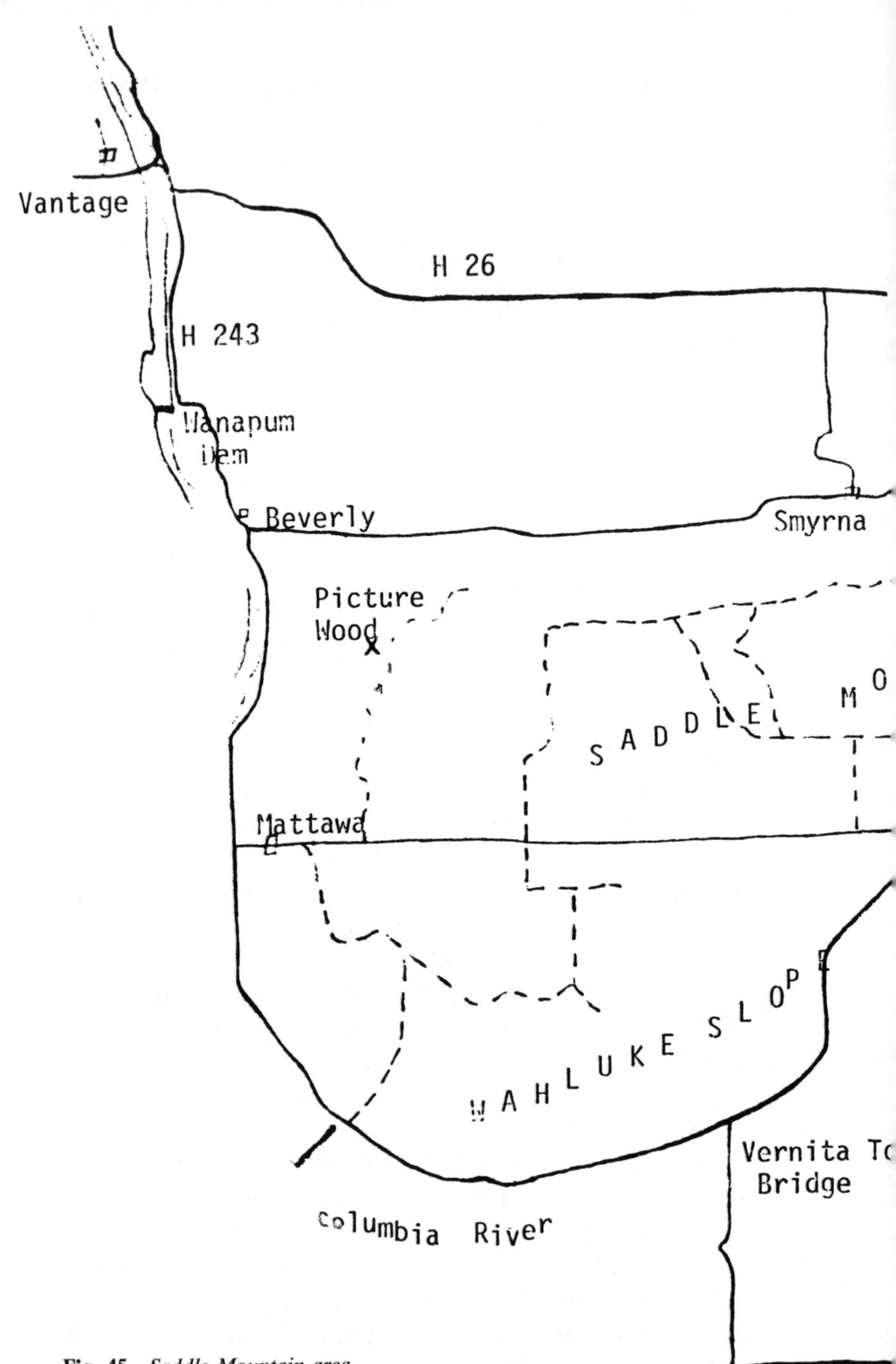

Fig. 45 *Saddle Mountain area*

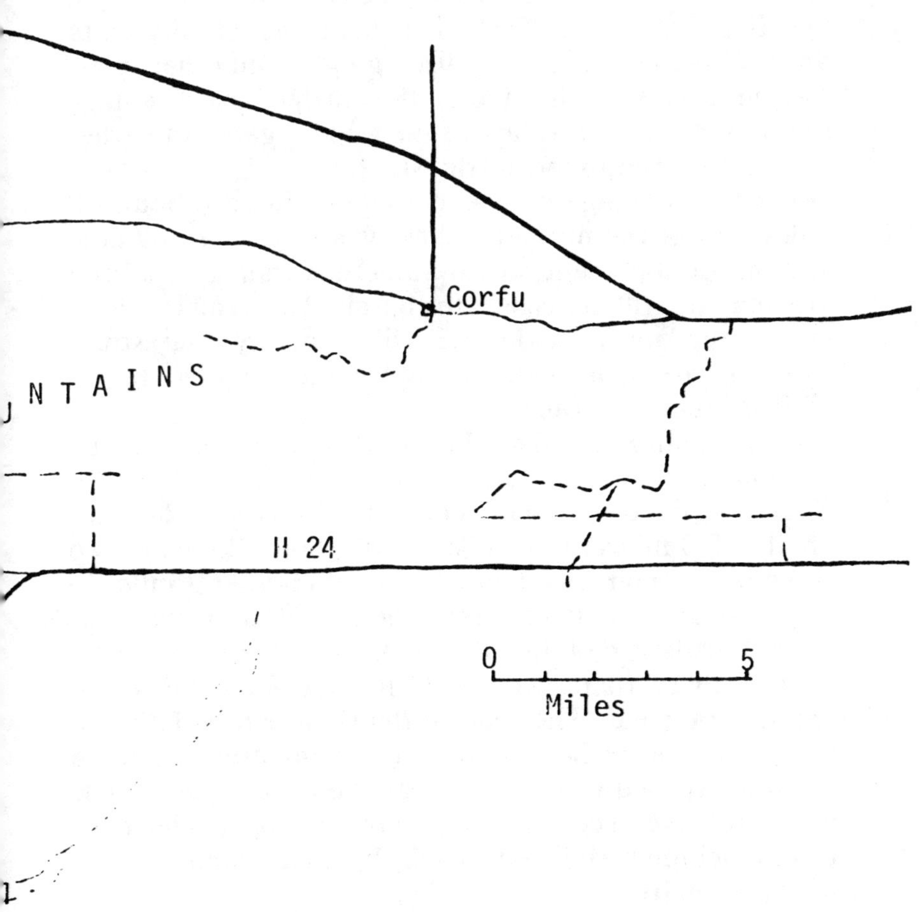

Corfu
NTAINS
II 24
0
5
Miles

mattock in the rocky soil, and a shovel will be useful to keep the the hole clean. On all public land there is a weight limit of 25 pounds and one piece of petrified wood per day per person.

West of Vantage there is a large area of basalt that contains petrified wood that extends into the Saddle Mountains. Much of the land is within the boundaries of Gingko Petrified Forest State Park. This park was established to preserve the deposits of opalized gingko and other trees. There is a museum located on the north edge of Vantage that has excellent displays of the wood, agates and other material that is found in the state.

West of Vantage on the Wanapum Road (about 2.8 miles west of the museum) there is a large parking area and information center with numerous trails to opalized logs that are still in place in the basalt. This is an excellent exhibit that most rockhounds will find very interesting. The trees are identified, and information can be obtained from the park ranger.

Good opalized wood can be found on the Game Department land that is located to the west of the park (Fig. 46). To reach this collecting area take the Wanapum Road for another 5.3 miles to the west. At this point there are two stone pillars marking a gate, proceed another 2 miles to a similar gate. This gate must be closed after entering. Several roads lead off in different directions inside the gate. Any of these roads will lead the anxious collector to deposits of wood. The road to the right can be followed to the ridge top where there is a parking area. Petrified wood is exposed in trenches near the parking area. The wood is of several colors of chalcedony and opal. The many colors include green, red, black, blue and combinations of these colors.

Large lots can be found by the rockhound who wants to work the basalts. Small pieces can be found on the surface by those rockhounds who are not in shape for the hard work that is required to "mine" the basalt. There is a 25 pound per day limit on the petrified wood on this land as on federal land.

Shovels are not very useful, except to clean broken basalt out of the hole. Most rockhounds use large bars, hammers, gads, chisels, and mattocks to collect petrified wood at this locality as in most of the localities in eastern Washington.

This is a very hot and dry area; be sure and carry adequate water when collecting anywhere in the central and eastern parts of the state during summer. Rattlesnakes can be found in some places, especially near the rivers. Usually they do not pose a threat, but all collectors should be aware of their possible danger.

An occurrence of opal, believed by some to be petrified wood, can be found east of Vantage near George. This deposit is reached by turning right off Interstate 90 on to Highway 243 after crossing the Columbia River at Vantage. Highway 243 is followed to Highway 26 where a left turn is necessary. Follow this highway to Route 281 where another left is made. This road is followed to the diatomaceious earth pits, where the opal is found.

The opal nodules are thrown aside as the diatomaceous earth is mined. Large pieces of opal can be collected in the waste along the pits. Collecting is allowed when the operators are not working.

Opalized wood is found along Umtanum Ridge south of Ellensburg. Wood can be found along the ridge and

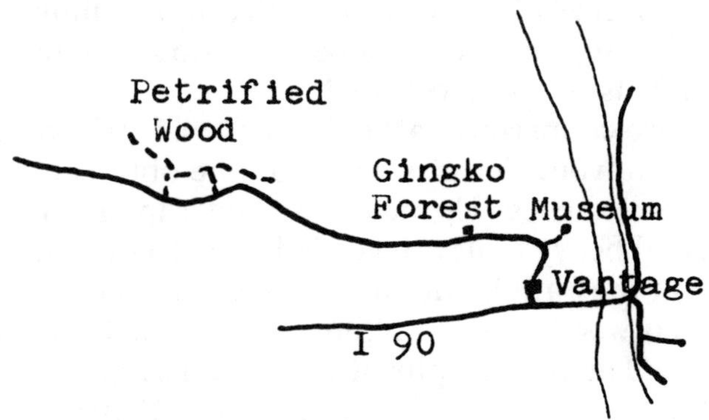

Fig. 46   *Vantage area*

along the dry canyons and creeks. The good localities are along Umtanum Creek and the two gullies to the south (Fig. 47). These are accessible via Highway 97, 8 miles south of Ellensburg (Fig. 48). The highway is along a very crooked section of the Yakima River Canyon. It is quite scenic and has numerous camping areas along the river.

A foot bridge gives access to the petrified wood deposits that are on the opposite side of the river from the highway. The bridge is visible from the highway when traveling south. A Game Department sign indicates the access to the bridge.

The wood is found along the length of the Umtanum Creek Canyon and the two canyons to the south. Opalized and agatized wood can both be found. Much of the wood is of the type that is known as bog. It is composed of a mass of roots, stems, limbs, vegetal matter and logs all petrified into a solid mass. Very good highly colored gem quality wood is found here. Figure 49 is of a pair of bookends that were made from Umtanum Creek wood.

There are numerous small canyons and side creeks along the Yakima River Canyon. Petrified wood can be found in the basalt outcrops in many of these canyons. Figure 50 shows one of these canyons with a group of rockhounds eagerly climbing up the canyon to a deposit of petrified wood.

The majority of these canyons are dry most of the year. The larger canyons that do carry water part of the year are good sources of pieces of petrified wood, limb casts, and bog. These small pieces of petrified wood can be found in the gravels of the creek bed.

There are numerous cattle ranches and private lands along the canyon. Permission should be obtained to collect on these ranches. The petrified wood is found in basalt, so the usual heavy tools are needed to collect in this area.

South of the Saddle Mountains there are good deposits of opalized wood in the Yakima Ridge and Rattlesnake Hills area. The topography of this area is similar to that of Umtanum Ridge and Yakima River Canyon. Some wood can be found by walking the dry washes. Most of

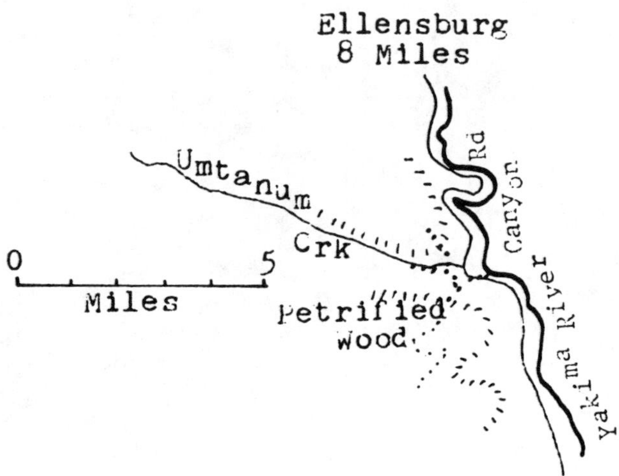

Fig. 47 *Umtanum Creek canyon, from the Yakima River canyon bridge.*

Fig. 48 *Yakima River Canyon-Umtanum Creek area*

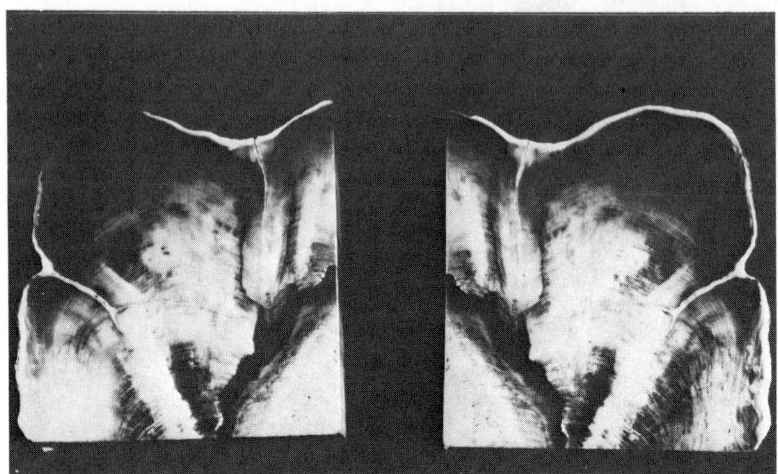

Fig. 49   *Petrified wood bookends, Umtanum Creek. Rudy Tschernich Collection.*

Fig. 50   *Members of the Skagit Rock and Gem Club searching for petrified wood, Yakima River canyon.*

the wood has to be removed from the basalt in several digging localities that are popular, but some pieces of wood can still be found lying free in the numerous gullies.

Local clubs and land owners can provide information about the best areas in which to dig. Be sure and close all gates that you open, and be courteous to the land owners. There are numerous wood deposits in this area and they will be opened for many years if all collectors obey the rockhound's code of ethics.

A fee area is run near the Silver Dollar Cafe south of the Vernita Toll Bridge. The cafe is located at the intersection of Highway 24 with Highway 241, 17 miles south of the bridge. Inquire at the cafe for directions and payment of the fee. The diggings are located about 3 miles from the cafe.

There is a deposit on each side of the highway at this locality. Digging conditions are the same in both areas. The usual hard basalt digging is required.

The wood is opalized and similar to that in Umtanum Canyon. Bog is quite abundant in large masses, and small limbs can be found along with large logs. Most of this wood is white in the center with brown streaks. The outside is brown or off-white. Some of the wood contains a few holes, but most of it is solid and takes a good polish.

This deposit is on public land, so the 25 pound limit is enforced. It is regularly bulldozed by the cafe owners, thus the fee, so that collectors do not have to dig through several feet of barren basalt to reach the wood.

Another area is located a few miles south on the Anderson Ranch. A $1 fee allows for 50 pounds of wood; additional wood is 2 cents per pound. The owners have bulldozed several areas to expose the wood. To reach this locality, go south on Highway 241 from the cafe for about 6 miles to the ranch house. The mailbox is marked Coby, or take Highway 241 from Sunnyside for 9.5 miles. Pay the fee at the ranch house and obtain directions to the diggings, which are about 2 miles north of the house.

The wood is mostly bog, as at the locality to the north. Large log sections and small limbs occur in the bog. Hard rock collecting tools are needed at this locality to remove the pertrified wood from the enclosing basalt.

Wood can be found in the area from Yakima to Moxee City to Sunnyside south of the Rattlesnake Hills. From here the petrified wood belt extends southwest to Bickleton, Goldendale and through the Horse Heaven Hills to Glenwood. To the east, petrified wood is found in the washes north of Highway 14 along the 15 miles stretch from Roosevelt to Alderdale. Some opalized wood has been found in the basalt south of Prosser and Mabton. East of this area there are reports of opalized wood in the Eltopia, Waitsburg and Dayton areas.

The area to the north of Vantage has produced some wood and is worthy of the same search as the region south of Vantage. Opalized wood has been found in Moses Coulee and Grand Coulee in Grant and Douglas counties. Large logs have been found in the basalts near Trinidad and gingko wood has been reported near Neppel.

There are deposits of petrified wood in several of the canyons and gullies east of Wenatchee. A good deposit is on the canyon rim east of Rock Island Dam. All of this region of eastern Washington is covered with basalt and possibly petrified wood. Inquire locally to find the best digging areas. There is ample room for many rockhounds who enjoy prospecting for new deposits, and many tons of wood to be collected in the numerous, already popular deposits.

Common Opal

Common opal of many colors can be found in the basalts and pillow basalts in central and eastern Washington. Large pieces of common opal have been found in construction sites in and around Spokane and Pullman. These opals are usually of a dirty brown to yellow color, although some are white and others light pastel shades. A few show moss-like inclusions of dark opal in light translucent opal. Few of the pieces of opal have attractive colors and most are full of holes and cracks. Much of this opal has a high water content. It cracks and falls apart as it loses this water.

In several road cuts between Rosalia and Spangle large masses of yellow to green opal can be found. This is the type of opal described above. Most of it is of poor quality, but because of the size of the individual masses, over 40 pounds, it does make interesting specimens.

This type of opal can be found in several areas in or near the petrified wood deposits. Very little of it is collected by rockhounds who are anxiously searching for that perfect specimen of petrified wood. A few rockhounds have discovered that the large pieces of the opal make good specimens for the rock garden.

Small pieces of common opal can be found with the agates in western Washington. Most of it is of a white or off-white color. Some pieces are banded and others are botryoidal. Some pieces are colorful and suitable for cutting a cabochon.

Precious Opal

There are several reports of precious opal in Washington, but the best material was found at Whelan near Pullman. High quality opal was discovered at Whelan near the Washington-Idaho border in 1890 (Brockett, 1974). The mines became known as the Moscow Opal Mines because of the interest shown by the people of that nearby Idaho community. There is little evidence that any opal was discovered on the Idaho side of the border.

For 4 years several opal mining companies came and went. The amount of opal produced is not known, and little accurate information can be found. Most of the literature quotes a figure of $5,000 worth (Huntting, 1960 and Dake, 1956). It is likely that this was the production of one company for only part of one year. It is possible that production may have been up to or more than $200,000 (Brockett, 1974).

The opal occurs in vesicles in Columbia River basalts. It was discovered during excavation for a well; after the first disovery, numerous occurrences were found on nearby farms. The opals were easily removed from the weathered basalt near the surface, but mining became difficult as the hard basalt was reached. Opals from the size of a pea to the size of an egg were found. The small opals had the best color, but a few large high quality opals were reported.

Most or all of the opal was of the white-base type. Color flashes were mostly of the pinfire or harlequin variety. At that time, the opals were reported to be as good as those from Hungary.

Presently the opal-producing area is producing wheat and the mines are filled in and closed. The owners of the property do not allow collecting by rockhounds. Good opal probably still occurs at this locality. It may not be commercially productive, but it may be good material for the collector. Perhaps some time in the future this opal may be "rediscovered" and opened to rockhounds.

Precious opal can be found near Moses on the Snake River (Fig. 4). This is above the green chalcedony occurrence that was previously described. This location is reached by taking the gravel road on the north side of the river from Lewiston, Idaho. The river is followed in a westerly direction for about 9 miles to where it turns north. There is a parking area along the road just before the corner, after the corner there is another parking area and a stile over the fence.

Access is easiest at the first parking area. A small canyon can be seen on the north. It is necessary to hike up this canyon for about 20 minutes to reach the opal. The canyon turns east, then north where it narrows. Shortly thereafter, it widens and another canyon branches off to the right. Go straight ahead, on the right side of the left branch of the canyon. A trail can be seen that goes up the grassy canyon side. Opal can be found in two pits about 5 to 10 minutes beyond the forks. If the pits are not found then cross to the other side of the narrow canyon and look back across the canyon to spot the pits.

Opal occurs in a gray, partially-altered basalt. Pieces of opal up to 1 inch across can be found, but most of it is about .25 inch across. Most of the pieces will be white to blue and some of it is brown. Some of the white opal will have layers that display a play of colors. This is mostly pinfire type and most useful for triplets.

Precious opal is not abundant, but some good small pieces can be found. The common opal pieces will tumble well if they are translucent. Other occurrences of this opal may be nearby. Larger workings have been rumored, but the author has been unable to locate them.

A hammer and chisel or gad will be needed to break pieces of basalt out of the pit. By carefully breaking these open, a few pieces of good precious opal may be found. Some material may be found in the basalt on the dump.

Rattlesnakes have been reported in this area, and black widow spiders can be found in their nests near the river. A careful eye should be open at all times, but there is no great danger. The biggest problem is the prickly pear cactus. On the steep canyon sides it is easy to put one's hands or knees on the long thorns of these cacti. They have a tendency to stick tenaciously to hiking boots and should be carefully removed.

There is abundant material in the basalts in the canyon. Small vesicle fillings of opal and chalcedony of many colors can be found by searching the talus of the numerous basalt cliffs in this area. A day's search will turn up enough good material to fill a tumbler.

Precious opal is reported with hyalite and hydrophane in vesicular basalt 2 miles north of Mondovi, 30 miles west of Spokane. The opal is reported to have a good play of color. Most of the pieces are small, less than .25 inch across. The opal is in layers, only a few layers will show color. A few excellent pieces have been reported, but they are scarce.

This is a wheat producing area near Davenport. There are numerous quarries and road cuts in basalt in the area. A thorough search of these basalt outcrops may yield a few pieces of precious opal.

Similar material has been reported from Moses Coulee. This is a dry gully that runs north-south about half way between Soap Lake and Waterville. There are good basalt outcrops along the coulee. Petrified wood may also be found.

There are thousands of basalt outcrops, road cuts and quarries throughout central and eastern Washington. Most of these have not been prospected by rockhounds. The author has found that most of these outcrops are of vesicular basalt. The vesicles are often filled with chalcedony and opal of many colors. Most of these are small, but the possibility of large pieces of good material being found in these outcrops is great. Some lucky collector may discover a small deposit of high quality precious opal in one of these outcrops. Prospecting the numerous exposures of basalt is an enjoyable way to spend an afternoon and often surprisingly rewarding.

RHODONITE

Rhodonite, a manganese silicate, is similar in use and appearance to thulite to the rockhound. It is a pink or white mineral, usually with black streaks, and is found in masses that are suitable for carvings, cabochons and bookends. Pieces that show a fine spider web pattern of black lines make unusual cabochons or slabs. This mineral has a hardness of 5.5 to 6 and polishes very well.

A small deposit of this mineral is found on Mt. Higgins, east of Mount Vernon. Excellent pieces of rhodonite have been removed from this deposit (Fig. 51). It occurs as a vein along the contact of a schist and serpentinite. The vein is exposed in a road cut alongside a rock quarry (Fig. 52). Numerous fractures and joints have broken the rhodonite into pieces from less than 1 pound to over 100 pounds. The pieces are coated with black manganese oxides.

This material can be collected by separating the pieces with a hammer and chisels or a bar. Each piece should be chipped to see that it is pink inside and not entirely black and porous. Much of it is white and of questionable gem value.

Access is difficult to this deposit. It is located on private timber land and reached via old logging roads. The deposit is located on a ridge top from which the view is of breathtaking beauty. One can behold Mt. Baker to the north and Whitehorse Mountain to the southeast standing in quiet dignity. Looking westward, one's eyes follow the contours of the Stillaguamish Valley and across the glistening Puget Sound before rising to the heights of the Olympic Mountains.

THULITE

The massive form of zoisite, known as thulite, has some appeal as a gemstone. Commonly pink, this material may also be white or green. Zoisite is a hyrdrous calcium aluminum silicate of the epidote group. Individual crystals have a hardness of 6, but it is usually found as a fibrous mass with a hardness of about 5. The material is best cut as a cabochon; large pieces may be carved or cut into bookends.

A few occurrences of thulite of gem quality are known in Washington. The best known deposit is along Tunk Creek, 5 miles northeast of Riverside (Fig. 53). Here the

Fig. 51 *Rhodonite, Mt. Higgins, bright pink. Mike and Gloria Johnson Collection.*

Fig. 52 *Collectors at the Mt. Higgins rhodonite deposit.*

thulite occurs in quartz veins in a hornblende schist. The color is pink, white and green. Most of it is solid, but some of it does not take a good polish. Pieces several pounds in weight may be found.

The material is found in a narrow canyon. A picture of the mouth of the canyon is illustrated in figure 54. Thulite can be found all along the creek from the mouth upstream for about a half mile. Search the canyon walls and boulders in the creek for pods of thulite and for quartz veins that may contain thulite. Some zeolites, tourmaline, actinolite and pectolite have been reported in these rocks too (Dake, 1956).

Thulite is reported at the Roosevelt mine near Republic. It is reported to occur with epidote, garnet and other contact metamorphic minerals. The occurrence is of small masses in the metamorphic rocks.

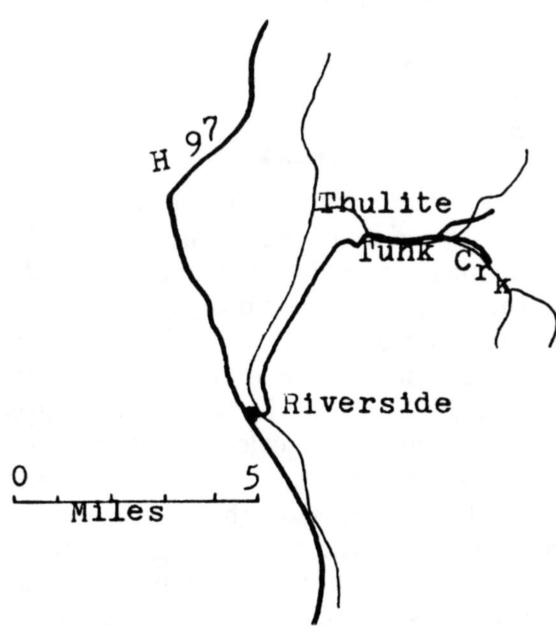

Fig. 53 *Tunk Creek area*

Fig. 54 *Tunk Creek, Thulite is found in numerous quartz veins in the canyon walls.*

ZEOLITES

Localities and occurrences of the zeolite group of minerals are described in the specimen section of this book. Because some massive zeolites can be cut into interesting cabochons they are briefly mentioned here. The zeolite group is a complex group of over 30 silicate minerals. Individually they form many habits, but the masses of needle-like crystals are of interest to the lapidary. Groups of terminated crystals of mesolite, scolecite, natrolite and thomsonite are of interest to mineral collectors; the masses of crystals are of little interest to them.

These minerals are quite common in volcanic rocks all over the world. Fortunately, Washington has a good supply of zeolites of many species. Cabochons that show a cat's-eye affect can be cut from almost any group of parallel or near parallel grown crystals. Several zeolites form this way and produce interesting white cat's-eye stones when cut. Mesolite from the Skookumchuck Dam was found in masses suitable for cutting stones over .5 inch in diameter.

This material is not available to the collector, because no collecting is allowed at the dam, but similar material may be found in the nearby basalts.

Interesting patterns can be formed if a stone is cut from material where crystals join from several directions (Fig. 55). This type of material is quite common in volcanic rocks that contain zeolites. Various species of the zeolite group of minerals occur in many areas in western Washington. Collectors should be able to collect some useful massive material at most zeolite localities.

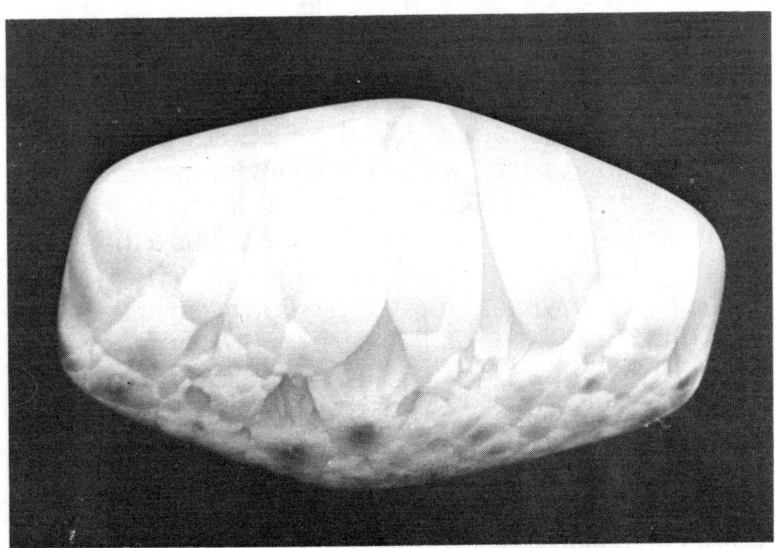

Fig. 55 *Natrolite cabochon, San Juan Islands. Ed Molsee Collection.*

PART II: MINERALS

ACTINOLITE AND TREMOLITE

The calcium magnesium iron silicate, actinolite, and the calcium magnesium silicate, tremolite, are the end members of a series. Tremolite, the iron-poor member, is generally white or a light color, as the amount of iron increases the color darkens and the mineral is called actinolite. Its color is generally green. These two minerals are members of the amphibole group and form monoclinic crystals which are elongated prismatic or are masses of acicular crystals.

They form under conditions of a moderate to high temperature and pressure, most often in a secondary environment in the presence of water. Both minerals may be found in granites and serpentines, gneisses, schists and marbles. Either mineral may form radiating coatings on joint surfaces or masses in the rock.

Pods of actinolite schist can be found at Lake Wenatchee. The schist is composed almost entirely of actinolite, muscovite and talc. Crystals of randomly oriented actinolite are abundant up to and sometimes over 2.25 inches long and .4 inch wide. By removing the muscovite and talc with small picks, excellent specimens of actinolite can be exposed (Fig. 56). This is a long and tedious job, but the resulting specimens are worth the effort.

The best known and most accessible occurrence of the actinolite pods is reportedly nearly worked out, but there are other areas nearby. The best area was east of Lake Wenatchee near Sears Creek (Lucas, 1974), but similar material can be found south of the lake above the Ranger Station (Fig. 56a). Good material can be found as float at the Sears Creek locality. This area is reached by following the highway past the lake across White River and onto the Sears Creek Road. Turn left onto the Sears Spur Road at the logged area and park where the road is blocked. Material can be found from the slide for about 2 miles up the road to the cliffs. Look for boulders of the actinolite schist in the rubble and brush of the hillside.

The other area is reached by taking the highway past the Ranger Station on the south side of the lake to a parking area about .5 mile east of the station. A short hike is required to reach the collecting area. There are other occurrences of actinolite in the Lake Wenatchee area. Permission to collect should be obtained at the Ranger Station. Collectors will need a large hammer and chisels to break the schist into pieces that are small enough to carry. Digging tools will be needed to remove some large boulders from the ground.

An interesting occurrence of stellate groups of tremolite crystals can be found in Illinois Basin on the North Fork of the Snoqualmie River (Fig. 56b). This area is mentioned and access described under epidote. The tremolite forms spheroidal groups in a peridotite (Fig. 57). Such an occurence is not unusual, but the size of the spheroids are. They are most numerous up to 10 inches in diameter, but several reach a diameter of 3 feet. These are abundant in the blocks of peridotite in the talus, but only small spheroids, less than 2 inches across, have been found in the peridotite in the cliff above the talus.

Weathering has exposed the tremolite in the peridotite, so they can be seen as light tan to silver-gray stellate masses in a reddish-brown matrix. Fresh tremolite varies from light gray to light green. Some of the tremolite is altered to talc.

The tremolite is abundant in the peridotite where the spheroids are exposed in cross section. It may not be possible to remove a fresh specimen from the matrix, but specimens with the weathered material exposed are easy to collect. They can be picked up loose or removed from boulders with a hammer and chisel.

These specimens are not showy or well formed tremolite crystals, but they are unusual. The occurrence and size

Fig. 56 *Actinolite, Lake Wenatchee. Specimen is nine inches across. Rudy Tschernich Collection.*

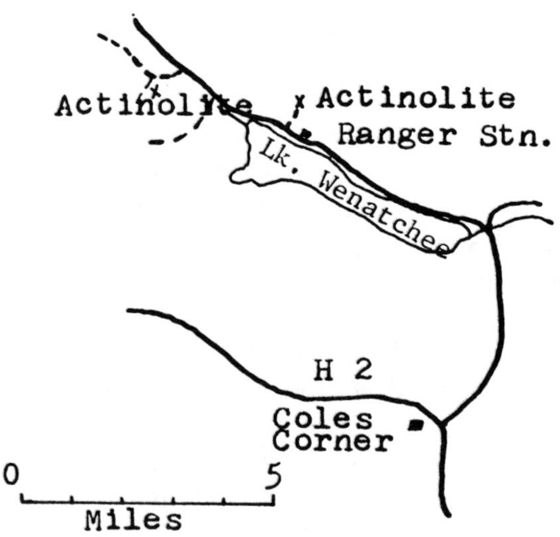

Fig. 56a *Lake Wenatchee area*

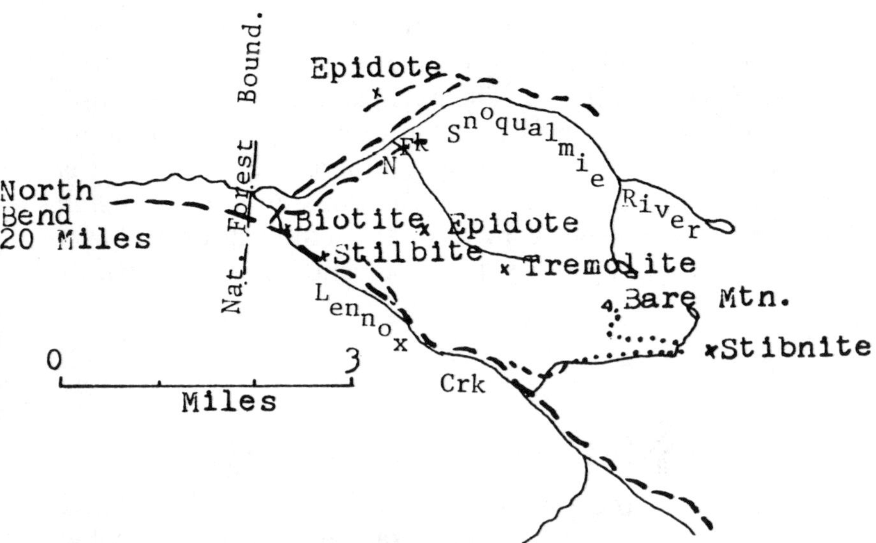

Fig. 56b *North Fork of the Snoqualmie River area.*

Fig. 57 *Tremolite in peridotite, Illinois Basin. (Mike Miller photo)*

of these spheroids are interesting to see. Unfortunately, the locality is difficult to reach as a long hike is required to reach the peridotite. The basin is brushy and swampy. During low water it is possible to walk up the boulders in the creek bottom as far as the ponds. From here it is necessary to walk in or along the talus on the south side of the basin. The peridotite is on the south side of the basin where the nearby flat valley floor suddenly steepens.

ANDALUSITE AND KYANITE

Andalusite and kyanite are two aluminum silicates that are often found together with staurolite in metamorphic rocks. Andalusite crystallizes in the orthorhombic system; crystals are usually square prisms terminated with the basal pinacoid. Kyanite crystallizes in the triclinic system and has crystals that are usually long and tabular. Andalusite is commonly various shades of red, reddish-brown and olive green. Kyanite is normally blue, white or green. Sillimanite, garnet, and corundum may be found with these two minerals.

Andalusite occurs with cassiterite, sillimanite, beryl, scheelite and wolframite at the Silver Hill Mine four miles south of Spokane. The andalusite occurs in Precambrian schists that are cut by Premesozoic and Mesozoic pegmatites, quartz veins, diorite and granite. The amount or quality of andalusite has not been reported. Except for a few quartz crystals this deposit has not produced many well-crystalized mineral specimens.

In Stevens County, andalusite is reported near Meadow Creek in the SE¼ of the SE¼ of sec. 35, T38N, R41E. Crystals up to 2 inches long have been reported. They are said to compose up to 50% of a schist near a granite contact (Huntting, 1960). Near the Longshot Mine andalusite is reported on a ridge that extends northeast from the Mine in secs. 7, 8, and 18, T36N, R41E. See update.

In Chelan County there are three locations where kyanite can be found. At the Railroad Creek location in the NW¼ of sec. 33, T31N, R16E, kyanite forms a .5 to 1.5 inch wide veinlet in Swakene gneiss. On Twenty-five Mile Creek in the W½ of sec. 1, T28N, R20E at an elevation of 3,400 feet kyanite crystals up to 2 inches long can be found in a mica schist. In the NW¼ of sec. 16, T27N, R18E, in Mad River Canyon on the east side of Maverick Peak there is an occurrence of a large mass of crystalline kyanite.

Andalusite is reported on Blakely Island in San Juan County. Crystals are found in the Wark gabbro where it is intruded by lamphophyre dikes. Andalusite has also been reported on Orcas, San Juan, Guemes and Fidalgo Islands.

In Skagit County, Kyanite has been reported at the Johnsburg Claim in the NE¼ of sec. 34, T35N, R13E. The crystals are found in a quartz vein in schist.

Much of the andalusite and kyanite occurs in schists with mica and quartz. Good specimens can be improved by carefully picking out the flakes of mica to expose the kyanite or andalusite crystals.

APOPHYLLITE

Apophyllite is a common mineral that is often found with members of the zeolite group. This potassium calcium silicate fluoride hydroxide hydrate crystallizes in the tetragonal system. Crystals are of well-formed prismatic habit, often of large size. They occur in several colors including colorless, white, pink and green.

Although it is found in few pegmatites, and some metallic mineral veins, its most common occurence is in volcanic rocks with zeolites. Generally, apophyllite occurs as well-formed crystals as singles and in groups of large size. It may occur in pockets alone or be associated with calcite, quartz or various members of the zeolite group.

Colorless to white crystals up to 1.5 inches long were found at the Skookumchuck Dam site. The crystals were of prismatic habit with pyramidal and basal pinacoid terminations. It was found in pockets in basalt with stilbite, mordenite, calcite, quartz and okenite. During excavation of the dam some excellent specimens were recovered. Collecting is no longer allowed at this locality.

Clusters of white prismatic apophyllite crystals can be found in the Pioneer quarry at Snohomish. The crystals occur in pockets in dark gray basalt. Specimens are not abundant, but a few good specimens can be recovered.

Pseudo-cuboctahedral crystals to 1.5 inch are found at the Pioneer Quarry south of Bremerton. Check at the quarry office weekdays for permission to collect.

ARAGONITE

The orthorhombic form of calcium carbonate is aragonite. Several crystal shapes are known for this mineral, but not as many as for calcite. Under surface conditions aragonite often reverts to calcite. The crystals will have the aragonite morphology, but the molecular arrangement will be that of calcite. The two minerals are difficult to distinguish from each other with normal physical methods.

There are few reported occurrences of aragonite in this state. Acicular crystals are found in the Columbia River basalts in several locations in eastern Washington. The best known may be in the area around Pullman. The occurrence of aragonite has been verified by X-ray analysis at the Geology Department at Washington State University. Two good localities were found by the author at Pullman (Fig. 58). The largest crystals were found at Busby south of Pullman. Smaller clusters of aragonite crystals can be found at the road cut on Highway 195 at the turn-

Fig. 58 *Argonite, Pullman. Vesicle diameter 5 mm.*

off to the cemetery on the west side of Pullman.

At both localities aragonite occurs as acicular crystals in radiating groups in vesicles in basalt. Other minerals that occur with the aragonite are described in the siderite section of this book. Most of the clusters are deposited on a layer of brown iron-opal. At the Busby locality many of the vesicles do not have a coating of the opal, but the aragonite reaches lengths over .5 inch. At the site near the cemetery the crystals are less than .25 inch in length. A few clusters of the aragonite have brown plates of iron-opal amongst the crystals.

The aragonite clusters are not common at either locality. It is necessary to break a quantity of basalt to find good specimens of aragonite and other minerals. It is likely that similar specimens can be found in the basalt outcrops elsewhere in eastern Washington.

Larger clusters of aragonite can be found in the basalt at Granite Point near Wawawai on the Snake River. Radiating clusters of aragonite up to 5 inches in length can be found in the scoriaceous zone at the top of the quarry. These are scarce, but a few specimens may be found by the ambitious collector. The vesicles in the scoria reach dimensions over 12 inches by 20 inches by 20 inches. This is a dangerous collecting locality. There are numerous loose rocks on the cliff. All collectors should wear hard hats at this site. Other minerals including pyrite and calcite may be found in this quarry.

ARSENOPYRITE

Arsenopyrite is a common gangue mineral in many metallic ore deposits. This iron arsenic sulfide crystallizes in the monoclinic system. The crystals are prismatic parallel to the "c" axis and are usually striated. The luster is metallic and the color silver-white. It is generally associated with silver and copper ores and with galena, pyrite, sphalerite, quartz and gold in hydrothermal vein deposits.

Small crystals are not uncommon in many of the mines of the state. Crystals may be found in vuggy veins and in vein material on the dumps. The author has collected crystals up to .4 inch across in several veins in the Silver Creek Mining District east of Index. The arsenopyrites are associated with small crystals of pyrite, galena, sphalerite and quartz. Most of the veins have matching walls with vuggy centers. Massive arsenopyrite is abundant on the walls of the veins and crystals are common in the centers. The veins are often filled with calcite. By dissolving the calcite with hydrochloric acid some good crystals may be exposed. The veins may be found in and around several of the mines in the district and in several of the road cuts.

Small crystals may be found in many of the other mines throughout the state. Most of the material will be massive but a few crystals may be found. The dumps of the mines will provide good collecting for arsenopyrite and other minerals.

AUTUNITE

Of the several species of uranium minerals, autunite is one of the most showy. It crystallizes in the tetragonal system; most crystals are tabular, usually in foliated and scaly aggregates. The crystals are yellow to dark green when fresh. This mineral is a calcium uranyl phosphate hydrate; it loses the water easily after being collected. The process is reversible and can also be prevented. Lost water can be replaced by putting the specimen in water and glycerine; unfortunately this process causes some autunite to soften and flake easily. To prevent the loss of water, coat the specimens with a clear sealant, such as acetate or lacquer.

Autunite that has lost water lightens in color; it is called meta-autunite I. If the specimen is heated to a temperature over $80^{\circ}C$, more water will be lost, and meta-autunite II is formed. The water cannot be restored after this state is reached.

Specimens of autunite are radioactive due to the uranium in their composition. In recent years, some doctors have reported that the radiation from radioactive mineral specimens may be dangerous to the collector that is exposed to the radiation of the specimens in his collection over a period of several years.

Spectacular specimens of autunite were collected at the Daybreak Mine north of Mt. Spokane. The autunite occurs as veinlets and fillings of joints in decomposed granitic rocks of Mesozoic age. Large crystal groups of excellent quality have been removed from some wide spaced joints and vugs. A few fine specimens were found that have been called pancakes. These are flattened masses of autunite with fine grained crystals on the outside and large coarse terminated crystal's on the inside.

The best specimens were of large free standing books of autunite or small plates. The books were commonly .5 inch across, with several books making up a good specimen. The plate-like specimens consisted of small books standing up like blades of grass on a joint surface. Most individual crystals in these specimens were less than .25 inch across.

Unfortunately for the relationship between miners and collectors, many specimens were stolen from the mine. The owners of the mine were interested in operating the mine for the production of uranium ore. They allowed very little specimen collecting by individuals, but did collect good specimens themselves. During operation of the pit a steady stream of collectors entered the mine before and after working hours and on weekends and removed what specimens they could. This is one of the reasons that collecting is no longer allowed at the locality.

Similar occurrences of autunite are reported at several nearby localities. These include the Dahl Uranium Mine, Dawn Uranium and Oil Co. Mine, and the Schafer prospect. None of these properties are currently producing or operating.

Good micro-crystals of autunite have been found at the Midnite Mine northwest of Spokane. They occur as individual crystals of .1 inch or less on matrix plates. The crystals are medium green and translucent.

Most of the uranium prospects in the state have produced some autunite. Collectors should obtain permission from owners before attempting to collect at any of these localities.

BARITE

Barite is a mineral that is common as a gangue in metallic veins, and in sedimentary rocks. It crystallizes in the orthorhombic system. Most crystals are tabular parallel to the base, but others are diamond shaped due to a vertical prism. Other shapes are not unusual. The color is white or colorless and light shades of yellow, red or blue. Specimens are easily recognized due to the high specific gravity 4.5, which is unusual for a nonmetallic mineral. The cleavage and crystal shape aid in the identification.

Barite deposits are located in several areas in Stevens County in northeastern Washington. Most of these deposits contain only massive barite, but a few contain well formed crystals. The best crystals are found in a vuggy zone along a fault in the Flagstaff Mountain deposit west of Northport. The mine lies at the top of Flagstaff Mountain on the south side. This deposit is massive bedded barite in Ordovician slates, argillites, limestones, and quartzites. Along a northeast trending fault at the front of an open pit, vugs contain smokey to clear crystals up to 3 inches long (Moen, 1964 and Mills, et al, 1971).

Some crystals are tabular but most are bipyramidal crystals elongated to the "a" axis (Fig. 59). Single crystals can be found broken loose from their matrix in the rubble around the fault. Careful searching and removal of material in the fault will expose many vugs lined with clusters of crystals. These crystals formed by recrystallization of the massive barite along the fault. Crystals are not known in other parts of the deposit.

Light to heavy tools can be used to collect these crystals. Most of the barite is broken up enough to allow removal of pieces along the fault with a small bar or chisel. Care should be taken as barite does have a good cleavage that makes crystals somewhat fragile.

Small tabular crystals of barite, less than .25 inch in length can be found at the Uribe Mine. The mine lies on the road 14 miles north of Colville. A small adit has been

Fig. 59 *Barite, Flagstaff Mountain. Cluster is 7.5 inches across.*

driven on the barite vein; the barite is fine to coarse-grained and no crystals can be found. There are small open joints and vugs in the argillaceous wall rocks on the east side of the adit. A hammer and chisel will be necessary to remove these specimens. Small doubly terminated milky quartz crystals and small galena cubes can be found in the argillite and quartz veins on the west side of the adit. The crystals are less than .75 inch in length and are scarce.

The Jacobson Mine, 11 miles north of Colville, contains some small tabular barite crystals. The crystals line vugs in the barite and are mostly less than .5 inch in length. Small euhedral pyrite grains suitable for micromounts can be found in stringers along the contact of the barite vein and rock wall. A heavy hammer will be necessary to break large pieces of the barite to find the vugs and remove the crystals.

Bright orange crystals can be found with dolomite in the Yellowhead Mine near Metaline Falls. The crystals are small, less than .15 inch on an edge. They occur in clusters in small vugs in limestone with sphalerite and dolomite. Several crystals may occur in a cluster on light gray saddle-shaped dolomite crystals. They vary from very light orange to bright orange. Excellent micromount specimens can be removed from the vuggy limestone at this mine. The Pend Oreille Mining Company has been doing exploratory work here. Permission should be requested from the company.

Large tabular barite crystals are rare with zeolites and quartz at Kosmos. Barite crystals over 1 inch on an edge and less than .25 inch in thickness have been found with quartz. A few fine specimens have been recovered, but are scarce now.

The first four deposits described above are on mining claims. They are not operating at this time, but permission should be obtained from the claim owners to collect specimens. There may be a rare occurence of euhedral crystals in some of the metal mines of the state. Crystals may be found on a few of the dumps of these mines.

BERYL

Beryl is a relatively common mineral of many colors. It is best known for its green and blue varieties that are known as emerald and aquamarine. Crystals of this hexagonal mineral are usually six sided prisms. It has a berylium aluminum silicate composition. Several rock types are hosts to beryl crystals including several types of metamorphic rocks and pegmatites.

There are no gem quality beryl deposits known in this state. In eastern Washington beryl is found in several pegmatites, but there are only vague reports of the mineral on the west side of the state. The pegmatites on Calispell Peak northeast of Chewelah have produced some good crystals. The Railway Dike pegmatite in the NW¼ SW¼ of sec. 33, T34N, R42E and the Calispell Peak pegmatite in the NE¼ NW¼ of sec. 21, T34N, R42E have produced beryl. Both of these pegmatites have a small history of production of beryllium ore.

There are eight trenches and an adit on the property of the Railway Dike. One large trench parallel to the pegmatite exposes the dike and mica schist country rock. Beryl crystals were found in the country rock side of the wall zone of the pegmatite. Some of the crystals are over 3 inches in length, but 1 inch crystals are most abundant.

The Calispell Peak pegmatite intrudes a portion of the Kaniksu batholith, a granitic body. The pegmatite is exposed in several trenches and large boulders at the mine. Beryl is found in the pegmatite along the border zone. As in the Railway Dike pegmatite, crystals are light green in color. Most crystals are less than 1.25 inches long (Fig. 60).

Both of these pegmatities are overgrown and well picked over. Collecting is poor, but a collector that is willing to break hard rock will find some fine crystals. A few large crystals over 3 inches long and 2 inches in diameter have been found in these deposits during the mining operations.

A few beryl crystals have been found in a pegmatite on the top of the ridge immediately south of the south slope of Mount Spokane. Exposure of the pegmatite is poor and much searching is required to find beryl. Crystals are reportedly less than 1 inch long.

Pegmatites are numerous on and near Granite Peak 8 miles south of Ione in Pend Oreille County. The pegmatites intrude quartz monzonite of the Cretaceous Kaniksu batholith. Crystals are reportedly green in color and up to 3 inches in length. The pegmatites can be reached by following the Ruby Creek road to the Triple H and J Mine road, then following this road to the pegmatites. From the

Fig. 60 *Beryl, Calispell Peak. Beryl crystal is .75 inch long.*

Fig. 61 *Aquamarine, Crystal Prospect. Crystal is 6mm in diameter. Gus Girnus Collection.*

end of the road at the pegmatites it is a long climb to the summit of Granite Peak. Beryl crystals can be found in several pegmatites at both locations. The mines are overgrown and the crystals are scarce.

Blue crystals of beryl are scarce in a pegmatite at the Kettle Falls Bridge on the Columbia River. They occur in a sill that is about 10 feet above normal water level. Presently, this sill is covered by the waters of Lake Roosevelt. At unusual times of extremely low water, access to this pegmatite is possible.

Beryl is reported in a pegmatite near Tunk Creek 4.2 miles northeast of Riverside. The pegmatite is exposed in a cliff that parallels the road. It intrudes a gneiss near a contact with Mesozoic granite. One aquamarine crystal was reported to have been found at this deposit (Pattee, *et al,* 1968).

Two small aquamarine crystals were found at the Crystal Prospect west of Springdale. The aquamarine crystals were found with smoky quartz, muscovite and orthoclase in a large vug in a small pegmatite. The two crystals were the only beryl crystals that were found with several hundred pounds of smoky quartz crystals. (Fig. 61). The crystals are light blue in color and not of gem quality; they contain inclusions near their centers. The largest crystal is .25 inch long and .25 inch diameter. Both crystals were well terminated.

Access to the deposit is difficult because of several side roads. Collecting may be done at this deposit, but the pegmatite is exposed on the surface of the ground and offers mostly hard rock to the collector. Small vugs that contain small smoky quartz and muscovite crystals may be found scattered throughout the deposit. In the future this deposit may provide an active collector with a few good specimens.

CALCITE

Calcite is a very common rock-forming mineral in sedimentary rocks and a common gangue mineral in ore deposits. This hexagonal rhombohedral calcium carbonate is found in many crystal shapes and colors. It is easily dissolved by ground-water and precipitated as stalagmites and stalactites in caves or as crystals in vuggy rocks and ore deposits. It is the chief constituent of marbles and limestones.

In Washington it is found in many localities, few of which offer particularly striking crystals. The Pend Oreille Mine north of Metaline Falls contains numerous vugs that are lined with sharp-edged acute rhombohedral crystals. Some of these are up to several inches across, but most crystals range from .5 to 1 inch. Most of the crystals are white and somewhat translucent. They are often associated with palygorskite and usually are found growing on sheets of this rare mineral (Fig. 62).

During mining operations vugs were encountered in the upper levels of the mine that were several feet across. Present mining operations at lower levels are uncovering very few calcite-filled vugs. No collecting is allowed at this mine.

Calcite can be collected in several of the mines and mining districts of the state. Small rhombohedral crystals can be located in a breccia in Red Gulch in the Silver Creek Mining District east of Index. This district was briefly mentioned under quartz. Acute rhombohedral crystals occur as coatings in vugs in the breccia. Most of the crystals are less than .25 inch across. Smaller crystals can be found in some of the numerous quartz and metallic mineral veins in this district.

Most of the mines and mining districts in the Cascade Mountains are in igneous rocks. Calcite is scarce in these mining districts, especially in large crystals. In the north-central and northeastern parts of the state sedimentary rocks are more abundant and calcite can be found in some

Fig. 62   *Calcite on Palygorskite, Pend Oreille Mine. Specimen is 4.4 inches long.*

of the mines in these two areas. Few large crystals will be found, but some small crystals and groups may be located on the mine dumps.

Acute rhombohedral crystals can be collected with quartz crystals in geodes in Walker Valley east of Mount Vernon. Most of the calcite crystals in these geodes are stained a light yellow by iron oxides. A few large clusters of calcite occur, but most of the crystals are less than .25 inch across. A few crystals will be found scattered across the quartz in the geodes.

Calcite is rare in the basalts of eastern Washington. Dogtooth crystals can be found with pyrite in the quarry at Granite Point on the Snake River. A description of this locality is given in the pyrite section. At this quarry the crystals are scarce; they are mostly less than .5 inch in length. A few crystals were coated with clay, the calcite has been dissolved leaving a hollow clay replica of the calcite crystals.

At the Monroe Quarry at Monroe, scalenohedral crystals to 4 inches occur in vugs in basalt. Crystals are white to pink in color and fluoresce a bright orange color. Hard rock tools are needed to collect these crystals.

At the Black River Quarry near Renton, yellow to white pyramidal crystals to 1.5 inches occur. The crystals show numerous parallel and stepped growths. Hard rock tools will be needed to collect the crystals at this locality.

Calcite crystals are common with the zeolites in the basalts of western Washington. Some large crystals were found at the Skookumchuck Dam locality and at Kosmos. Most of the other localities produced calcite crystals of small size.

CHALCOPYRITE

Chalcopyrite crystallizes in the tetragonal system but euhedral crystals are rare. This brass-yellow copper iron sulifide is generally found as granular masses. The crystals and masses generally tarnish to an irridescent rainbow of color. Crystals are commonly disphenoids and scaleno-hedrons.

This mineral is very common in the numerous metalic mineral mines of the state; it is associated with pyrite, galena, sphalerite, and quartz. Few crystals have been reported in the state. It is found in nearly every mining district, so it is likely that a few crystals may occur. Search the dumps of the old mines for small crystals of micro-mount size.

Large crystal masses showing crude faces have been found with quartz and pyrite at Spruce Ridge. These have been reported to be up to 1.24 inches across. Only a few specimens have been collected.

Good tetrahedral crystals to 1 inch can be found in the Pedro Pipe on the south side of Katy Bell Ridge with quartz, pyrite and galena. Access to this deposit is very difficult and is described in the quartz section.

Tetrahedral crystals to 1 inch occur in vugs in an altered volcanic breccia in the Mono Mine on the Miller River. Crystals are a bright brassy color and coated by small sphalerite crystals. Quartz and ankerite crystals are associated. The rocks vary from green to gray to white and are generally hard. Collecting requires hard rock tools.

COPPER

Native copper crystallizes in the isometric system as do the other native metals. Crystals are tetrahexahedrons, cubes, dodecahedrons or octahedrons. Most crystals are malformed and in branching arborescent groups. Most common are masses, plates and scales, or twisted wire-like forms. Copper may be found concentrated in stream gravels because of its high specific gravity, 8.9.

Copper has been found in most types of rocks, in the oxidized portions of sulfide copper deposits or in copper veins. Native copper is reported in many extrusive igneous rocks. In these basalts, andesites and rhyolites, it is found disseminated in the rocks, in veins or as coatings on joint surfaces.

Native copper is reported in several localities in this state. The most common occurrences are traces of disseminated copper in the host rock of copper deposits. Unfortunately these occurrences are of little interest to collectors.

Traces of native copper have been reported from the manganese deposits of the Olympic Peninsula. Various manganese minerals are found in Eocene volcanic rocks.

The copper is found as plates and scales in joints in the rocks, or in the manganese oxides. The deposits are found on the north, west, and east sides of the peninsula. The Lake Crescent locality is noted for manganese and some copper. No mines however, are noted for an abundance of copper specimens.

Near the Skookumchuck Dam zeolite locality, copper has been found in Eocene basalts. This occurrence is near the confluence of a small side stream with the Skookumchuck River about three fourths of a mile above the dam. The reservoir above the dam has covered this occurrence. The copper is found as sheets and plates in joints in the basalt.

DIAMONDS

The beauty and value of a diamond is known to everyone. The rarity of this mineral has made it the most difficult mineral to be collected by amateur collectors. Diamonds are the isometric form of crystallized carbon, and are the hardest natural substance known on earth. The hardness of 10 makes it ideal as a nearly indestructible gem and preserves its shape as crystals are carried in the gravels of a stream where they are concentrated in placer deposits.

Diamonds occur most abundantly in a rare peridotite known as kimberlite. No kimberlites are known in this state but other peridotites are common. Placer deposits being worked for gold have yielded a few diamond crystals. Most of these finds of this rare mineral have been poorly documented, and the majority are probably erroneous.

A 4 carat yellow diamond found in a Skamania County gold placer in 1932 is probably the largest diamond recovered in this state. This find is not documented, but there are enough sketchy reports about this find to indicate that it is probably true. A diamond of this size would not be too difficult to spot in a gold placering operation, but most diamonds in the placers are less than one quarter carat in weight. Diamonds this small would be more difficult to see amongst the pieces of quartz and other minerals in a gold concentrate.

More than seven diamonds have been reported from the old placers of Kittitas County. The amount of gold placering operations and the amount of peridotite that occur in this county make it probable that several small diamonds would be found.

Microsize diamonds have been reported from the black sands of the coast. These occur with magnetite and ilmenite in the sands of the ocean beaches.

A few diamonds have reportedly been found along the Sauk River. There are a few peridotites in this area that may be the source of these diamonds. It is probable that the diamonds would be of a micro-size, mostly less than .1 carat in weight.

Anyone interested in searching for diamonds and possibly a diamond pipe needs to do an extensive program of carefully concentrating gravels and searching for diamonds and the minerals that are generally associated with diamonds. These indicator minerals are pyrope, chrome-diopside and magnesium-ilmenite. The average mineral collector could do some prospecting for diamonds without any more work than panning for gold. All that is needed is a shovel, gold pan and hand lens of 10 power or more. The method of panning would have to be more refined than that for gold, because the specific gravity of diamond is only 3.52, not much greater than that of the gravels that contain the small crystals. Search the concentrates in the pan carefully by eye then with a hand lens for micro-size crystals and pick out any crystals and gold flakes with a wet artist's brush or tweezers. More minerals, including other micromount size crystals may be found and recovered if the concentrate is saved and perused under a microscope.

The best areas to search would be the streams that flow

off of the central and northern Cascades. The areas and streams where serpentine and jade can be found may be good areas, because of the associated peridotites. A collecting trip for diamonds will be hard work, but possibly rewarding.

EPIDOTE

Epidote is a common mineral that is usually found as masses in igneous and metamorphic rocks and in ore deposits. The calcium aluminum silicate crystallizes in the monoclinic system. Excellent long, grooved prisms are common, but smaller radiating acicular groups and masses are more so. The long prisms are usually elongated horizontally so that the basal cleavage is vertical when a long crystal is set on one end. Normally, epidote is light to dark green, but brown or yellow-green epidote does occur.

Excellent crystals have been found in contact metamorphosed limestones and related skarn zones. Crystals are often found with zeolites in trap rocks, but the mineral is most common as coatings or masses in joints and cracks in igneous and metamorphic rocks.

Well-formed prismatic crystals of large size have been found in Washington. The best locality is probably Denny Mountain. Long prismatic crystals are found in calcite filled vugs with clinozoisite, pyrite, quartz, diopside and grossular. The occurrence is in a skarn in a steep narrow canyon. Epidote, dark green in color, forms prisms up to 2 inches in length and .6 inch in cross section. They occur in groups that extend across or radiate into the calcite filled vugs as sheaf-like crystal clusters. Clinozoisite has a similar occurence at this locality and is difficult to distinguish from the epidote.

By removing the vugs and etching out the calcite, fine

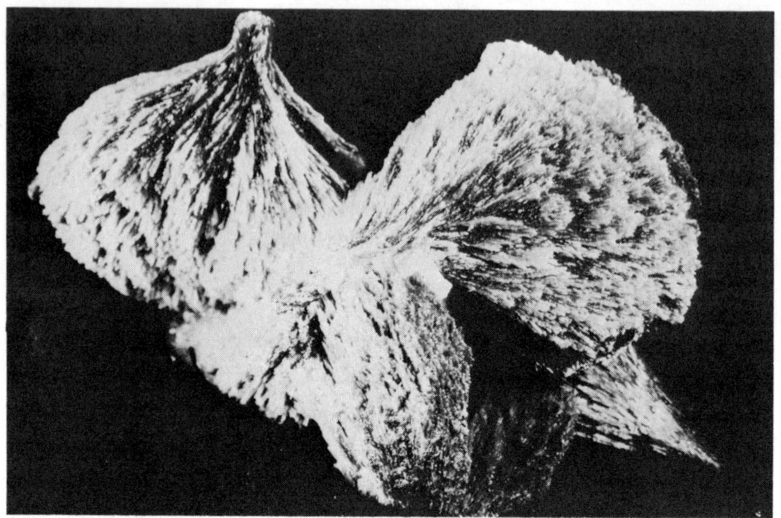

Fig. 63 *Epidote, Sunset Falls. Cluster is 3.75 inches long. Rudy Tschernich Collection.*

specimens of epidote with grossular, quartz, specular hematite and pyrite can be obtained. Specimens can be found in the canyon walls or talus. Heavy tools are needed to remove specimens from the walls.

A few excellent crystals of epidote were collected in vugs in diorite in a railway cut paralleling Highway 2 near Sunset Falls, about four miles west of Baring. Similar material is reported to occur with garnets and mica on the ridge between Barclay and Trout creek north of this locality.

The epidote occurs in radiating groups. Specimens with albite and quartz were collected a few years ago (Fig. 63).

Specimens of similar habit may be found in the Illinois Basin on the North Fork of the Snoqualmie River Small flattened clusters of radiating acicular crystal habit to 1.5 inches long have been found associated with quartz in the abundant talus slopes of the west side of the basin. A combined trip for the epidote and the tremolite which occurs further up the basin might be considered.

There is no trail into the basin at present as the old mining trail is well overgrown. Access is easiest up the creek when the water is low. Stay to the left of the falls and climb up the bushy hillside to the lip of the basin. Above the falls stay on the right side of the basin. The first part of the basin is swampy and brushy. About one hour is required to hike from the road to an old cabin. From the cabin a fifteen minute hike is needed to get to the gulches and talus to the south and west of the cabin. See update.

FELDSPARS

The feldspar group of minerals includes several individual species. Except for some crystals these are usually difficult to distinguish from each other. These are various potassium sodium (orthoclase group) or calcium (plagioclase group) aluminosilicates that crystallize in the monoclinic and triclinic crystal systems. The orthoclase group, except for microcline, crystallizes in the monoclinic system and the plagioclase group in the triclinic system. The colors of the minerals vary from white to pink, gray to dark blue, and sometimes colorless.

Feldspars are abundant as rock-forming minerals. In some granite rocks large phenocrysts of orthoclase may weather out where they can be collected. Such occurrences have been reported in some granite rocks of Stevens County, and possibly in the Cascade Mountains.

Large, light pink crystals of orthoclase, over 4 inches long have been found with smoky quartz, muscovite and beryl at the Crystal prospect west of Springdale. A few specimens have been recovered with quartz crystals (Fig. 61). They occur in vugs in a pegmatite dike; the deposit is described under quartz and beryl.

Small crystals can be recovered from basalt vesicles in eastern Washington and possibly western Washington too. Most crystals are less than .1 inch in length. These are usually white to clear and are generally associated with several other minerals such as magnetite, cristobalite and siderite. A feldspar varnish formed by a thin coating of crystals occurs in many of the vesicles. A few good micromount specimens of these feldspar crystals may occur in any vesicular basalts of the state. The basalts and minerals of eastern Washington have been described in the siderite chapter.

Crystals of albite occur in a pegmatite in a roadcut on the Icicle Creek Road 4 miles southwest of Leavenworth. The albite occurs in white tabular crystals up to .6 inch long intergrown with chlorite. Masses of rose quartz and crystals of beryl occur at this locality.

White crystals of albite under .5 inch occur with quartz, tourmaline and chlorite in miarolitic cavities in the granodiorite in Bear Basin (see the access description in the stibnite section). Most cavities are under 2 inches across. The quartz crystals are generally colorless and less than .5 inch long. Tourmaline occurs as black radiating crystals or rarely as short prisms. Specimens can be collected in several areas in the abundant talus on the east and south sides of the basin. Collecting may not be as easy or as good in the abundant outcrops in the basin. Tools for hard rock collecting are required.

Microcline occurs with riebeckite, astrophyllite, quartz and zircon in miarolitic cavities in the Golden Horn batholith in the area around Washington Pass on the North Cascades Highway. Cavities containing crystals are scattered and can be found at several localities in the area. Hard rock tools are needed for collecting these fine specimens.

Microcline crystals to .75 inch diameter occur in miarolitic cavities in the diorites around Lake Serene south of Index. The crystals occur with penninite and quartz of the Tessin habit. A difficult hike is required to reach this locality and hard rock collecting is the rule. See update.

GALENA

The lead sulfide, galena, is more common in the state as crystals than either sphalerite or chalcopyrite, with which it commonly occurs. Crystals are rarely anything but cubes; although octahedrons occur. The perfect cubic cleavage, natural cubic crystals, and gray color along with the high specific gravity make the mineral easily recognizable.

Large crystals are scarce in this state, but small crystals up to 1 inch across can be found in several mines. The Big 9 Mine in the Orient District of Stevens County is a source of crystals up to 1 inch on an edge. These crystals can be found on the dump or in the vein of quartz on the deposit. Some crystals in the quartz have a length about three times their width and are curved, making rare and interesting specimens. This unusual shape is probably due to deformation of the deposit. Some free cubes and cubes in matrix can also be found.

About one mile west of the Big 9 Mine there is an occurrence of galena with quartz. The galena cubes occur in groups with small quartz crystals. The cubes and quartz crystals are intergrown in branching groups, some of which contain a few small pyrite cubes. These specimens are rare. There is little mineralization at this deposit; only minor prospecting has been done. A few specimens may still occur on the small dumps. The country rocks are Permian sedimentary rocks; nearby are Mesozoic volcanic rocks.

Small galena crystals have been found in several mines in the Covada Mining District west of Incheleum. This district covers nine square miles around Covada Lake in the NE¼ of T32N, R36E. Small galena cubes can be found in the gossan on the mine dumps. There are also small crystals of sylvanite, pyrargyrite, wire silver and gold in the quartz on the dumps.

Small flakes of gold have been found in the quartz and in vugs in the quartz. Search the dumps throughout the district, mineralization is similar in all the mines. The mineralization is found in the quartz veins that cut granodiorite and pyroxenite.

Unusual stalactitic growths of galena are rare with quartz, pyrite and chalcopyrite at the Pedro Pipe breccia on Katy Bell Ridge along the Middle Fork of the Snoqualmie River. The stalactites reach a length of 1.5 inches.

GARNETS

The garnet group contains six common species of minerals and eight rare species. The six common species have various compositions of magnesium, iron, manganese, or calcium-aluminum, calcium-iron or calcium-chromium silicate. New species are being discovered with titanium or rare earths substituting in the formulas for the common elements.

All species of this group crystallize in the isometric crystal system. Crystals are commonly well-formed and usually display a dodecahedron or trapezohedron shape. Combinations of these two forms are common; cubic crystals are rare.

Garnets are found in metamorphic rocks and a few igneous rocks. They are most abundant in schists and gneisses and are common in pegmatites. Because of their hardness and specific gravity, 3.5 to 4.3, they are common constituents of placer deposits. The crystal shape is often maintained with little abrasion from the transportation with the softer minerals of the stream sediments.

A vitreous luster and sharp distinct crystal shape adds to the beauty of the matrix specimens. Crystals of large size, several inches in diameter, are common in some localities. The contrast of the garnet with the white, black or gray matrix helps make the garnet specimens pleasing cabinet material.

Vesper Peak in Snohomish County is the most noted garnet-producing locality in Washington. The deposit is on the northwest side of Vesper Peak at an elevation of 5,500 feet. Deep winter snow makes collecting impossible except for a short period in late summer.

The deposit has been held by mining claim since the 1950's. A succession of owners have tried with limited success to make the deposit commercially viable. The current owner, Bart Cannon of Seattle, mined many fine specimens in 1981 and 1982, and is continuing working. Mr. Cannon does not permit unauthorized collecting.

The grossular garnets, variety hessonite, occur in a vuggy carbonate skarn with quartz and diopside. Most of the garnets are .25 to .5 inch in diameter and are a light orange to dark brown color. Most of them are translucent, a few are quite clear. They are dodecahedrons and trapezohedrons, with cube faces on a few specimens. The cube faces have a dimpled surface while the other faces are smooth (Fig. 64).

They can be collected as good singles, but mostly they are good matrix specimens. A few specimens have been found that have quartz crystals with the garnets. The quartz is milky and well terminated. Diopside is scarce, but it can be found as brown acicular crystals. Few attractive specimens of diopside have been collected, but they are interesting specimens.

Grossular crystals can be collected on Denny Mountain. These crystals are similar to those on Vesper Peak, except they tend to be more brown than orange, and are not as translucent. The crystals can be found in two localities. They have been recovered from the skarn zone at the quartz crystal deposit; this area has been thoroughly described in the quartz chapter. At this locality the crystals average about .25 inch in diameter. They can be found lining vugs in carbonate and massive magnetite rock. Most of the vugs are filled with white calcite, and some of them contain dark green epidote crystals. Good specimens can be recovered, but hard rock must be broken to get them.

Doubly terminated Quartz cluster

Fig. 64 *Grossular and quartz. Vesper Peak. Largest garnet is .5 inch. Bob Jackson Collection.*

116

The other locality is of a similar occurrence, located in the lower part of Denny Creek Canyon. The deposit can be reached by hiking up the Melakwa Lake Trail for about 15 minutes from the parking area near the Denny Creek Campground. At this point, cut through the woods to the left to Denny Creek, where there is a falls and narrow canyon. The garnets can be found in veins and vugs in the rock above and below the falls. This carbonate rock is similar to that at the other deposit on Denny Mountain. Epidote and quartz are more scarce, but do occur in the vugs. The rock above the falls is very rounded and difficult to collect in.

The best collecting may be in the canyon below the falls. The sides of the canyon are nearly vertical, so the best way to get in the canyon is to hike up the creek from the foot bridge, where the trail crosses the creek below the falls. About a 10 to 15 minute hike is required from the bridge. Because of high water in the creek, collecting is only possible from about July to September. The canyon is very narrow and collecting is difficult even when the water is low.

The garnets here are larger than those at the deposit in Rockhound Gulch. Dodechedral crystals up to 1 inch in diameter can be found in vugs in the rock walls of the canyon and in boulders in the creek below the falls. This rock is hard and difficult to break. Good specimens can be obtained by taking the calcite vugs home and etching away the calcite with hydrochloric acid to expose the crystals.

Garnets have been reported one mile south of Cascade Pass east of Marblemount. A long drive and hike is required to reach this locality. The garnets are reported to be pyrope and up to .75 inch in diameter (Huntting, 1960). They are found in the greenschist of Prejurrassic age. Very little is known about this deposit, but good material may occur here. It is unlikely that the garnets are pyrope, but they may be very good specimens.

This locality is in the North Cascades National Park. If the status of this land is changed to wilderness classification, collecting will be allowed.

Andradite occurs as dark olive colored crystals to .5 inch diameter at Green Mountain Summit near North Bend. The occurrence is in a small skarn and associated minerals include specular hematite, magnetite and axinite.

Crystals of andradite to .6 inch occur along limestone-diorite contacts with pyrite, magnetite, malachite and diopside in the quarries at Moloney Mountain near Skykomish. Crystals are diploidal and are a yellow to olive green and red in color.

At the Dutch John Mine near Carlton, andradite crystals of a red brown to olive green color occur in a skarn. These dodecahedral crystals reach a diameter of .9 inch and are associated with scheelite, quartz and magnetite.

Grossular crystals occur at the Guye Peak Saddle north of Snoqualmie Pass. Crystals to 3.5 inches have been reported and are brownish red and semitranslucent. They occur with magnetite and epidote. The occurrence is in a contact zone of the Snoqualmie batholith with sediments of Pretertiary age.

There are numerous locations in Washington where metamorphic rocks occur; these are potential sources of garnets. A review of the geologic map of the state will indicate several possible places to search for garnets.

GAYLUSSITE

Saline Lakes are known to produce several sodium, calcium and borax carbonates and hydroxides. Many of these form good crystals and are quite rare. Gaylussite is a hydrous sodium calcium carbonate. Euhedral crystals are rare, the habit is usually massive to granular. In Washington, gaylussite can be found in Mitchell Lake located 1 mile southwest of the Wilson Creek station on the Burlington Northern Railway in Grant County.

The crystals are common up to .5 cm across. Most crystals are water-clear; they are elongated on the "a" axis

(Bennett, 1962). Crystals occur disseminated in a 2 foot thick zone of mud in the middle of the lake at a depth of 7.5 feet. Recovery is difficult and is easiest in late summer and early fall at low water.

GEOTHITE
(Pseudomorphs of Pyrite and Siderite)

Pyrite and siderite are common minerals in all types of rocks, but they are not stable under normal atmospheric conditions, especially in the presence of water. The iron sulfide and iron carbonate is easily replaced by the iron oxide hydroxide, goethite. The resulting pseudomorphs have the pyrite or siderite crystal shape but the composition of goethite. Such pseudomorphs are usually dark brown to black.

These pseudomorphs are quite common in rocks that contain pyrite or siderite. Pseudomorphs of pyrite are especially abundant in sediments or metasediments. In Washington good goethite pseudomorphs occur at Tekoa Mountain south of Spokane and at Denny Mountain. At Tekoa Mountain (Fig. 65) the crystals occur in a light-gray to light-brown to Precambrian quartzite. There is no best place to collect on the mountain. In the past the crystals were quite common on the surface, but active collecting has removed most surface material.

There are numerous roads on the mountain; the road cuts offer the best collecting. Crystals are collected by breaking pieces of quartzite along the bedding planes. Crystals will be found as singles and groups of crystals (Fig. 66). Hard rock tools will be needed to collect these unusual specimens.

In the past, large crystals up to 2 inches on a side were not uncommon. These are rare now as the abundant surface material has been removed. Most crystals will average around .5 inch, but those that are 1 inch on an edge can often be found. Most of the crystals can be collected in matrix; a few specimens are loose and will fall out of the quartzite.

Apparently the crystals are concentrated in beds, possibly in groups. If one crystal is found it is likely that a few more will be found in the immediate vicinity. A collector may be lucky and find several fine, large crystals, or be less fortunate and find only a few medium-sized crystals in a days collecting time.

Similar crystals can be found on Denny Mountain with the large quartz crystals above the magnetite zone and within the magnetite. Pseudomorphs over 2 inches on a side have been collected with quartz crystals. These often occured in clusters of several goethite pseudomorphs with several quartz crystals. A few small specimens may be found in the small vugs as all large exposed vugs are cleaned out; hard work will expose more material.

The pseudomorphs vary from complete to only partial replacement. Fresh pyrite may show through some of the brown crystals. If cleaned in oxalic acid the goethite can be removed exposing the pyrite. Most of these specimens will then have a rough surface.

Goethite pseudomorphs of pyrite are not uncommon in this state. Except for the two localities described above the crystals are usually less than .5 inch on an edge. In many metamorphic and sedimentary rocks and a few igneous rocks, small pseudomorphs, less than .05 inch are numerous. In these localities a few larger specimens may occur. It is worth the collector's time to check these occurrences for an occasional fine specimen.

Several localities for siderite were described in a preceding chapter. The siderite in basalt vesicles in eastern Washington is commonly replaced by goethite. These pseudomorphs are usually black or brown; the siderite was light brown, olive or reddish-brown in color. In most specimens the alteration to goethite can be detected by the color or the opacity of the crystals. Most of the siderite is translucent. Some fine thumbnail and microsize specimens of these pseudomorphs can be collected with the siderite.

GOLD

Gold is considered to be quite rare, but it can actually be found over much of the world in all types of rocks. In the native state it is found as nuggets in stream beds or as nuggets or crystalline groups in quartz, calcite and other host rocks. Nuggets are the most common form of native gold; these may have any shape from flat flakes to nearly round marbles. They have been found in sizes up to 190 pounds or down to pieces that are so small that over 14 million are needed to make an ounce.

Crystalline gold belongs to the isometric system. Crystals are commonly octohedral; most often are distorted and attached together to form arborescent groups, dendrites, or filiform groups. If such specimens are recovered in situ or from the stream near their source the shapes will be maintained, but they become rounded as the gold is roll-ed and pounded as it is carried downstream.

The high specific gravity, 19.3 if pure, of gold causes it to settle rapidly in the gravels of a stream and be con-centrated on the rockbed. This high specific gravity also allows for the easy recovery of the gold from the gravels by panning and sluicing operations.

Most gold is not pure in the native state; it is naturally alloyed with silver. If the amount of silver exceeds 20% the alloy is called electrum. Other metals may be present including copper, iron or platinum. It is common to recover nuggets of silver, copper or platinum from the same stream with the gold.

Gold can be recovered from the streams that flow off of the Cascade Mountains and streams in parts of the northeastern and northwestern parts of the state. Swauk

Fig. 65 *Tekoa Mountain as seen from the highway south of Tekoa.*

Fig. 66 *Goethite pseudomorphs after pyrite, Tekoa Mountain. Center crystal is .6 inch across.*

Creek, northeast of Cle Elum, is the most noted occurrence of crystalline gold in Washington. The gold was first recovered as wire gold and nuggets from the creek. Later it was discovered in lode deposits from which excellent arborescent groups of gold have been collected (Fig. 67). In the lode deposits it occurs in pockets and veins in quartz and calcite along shear zones. The best crystalline gold specimens from the state have been mined from these pockets in recent years.

Few mines are being worked now, but some gold is recovered every few years. Some miners are willing to sell the gold for mineral specimens. The creek has been dredged and side creeks have been heavily worked. Some small gold specimens can be recovered by working the gravels. Search the windrows of gravels that were left by the dredge; reports indicate that the screens in the dredge were small and that large nuggets went out with the gravels. Nuggets have been found that weigh up to 50 ounces. Most of this ground is claimed, so collectors should inquire before panning for specimens.

Peshashatin Creek, over the pass to the north of Swauk Creek, has a similar history to that of Swauk Creek. Nuggets were generally smaller in this creek and the gold contained more silver than that in Swauk Creek. There are numerous side creeks that contain gold. This area offers the collector numerous places to collect gold and agates and much scenery to enjoy.

Most streams in washington will produce a little gold to the collector who is willing to work for it. Several of these streams have produced gold in paying quantities and others contain gold, but in quantities that are not economic to recover. These streams can produce several flakes and small nuggets to the weekend prospector who is not afraid to work. The streams in the Okanogan area and gravel bars of the Columbia and Snake Rivers have produced fine gold.

Rare flakes of gold have been found in the rocks in the mines of the Covada Mining District. Specimens are not plentiful.

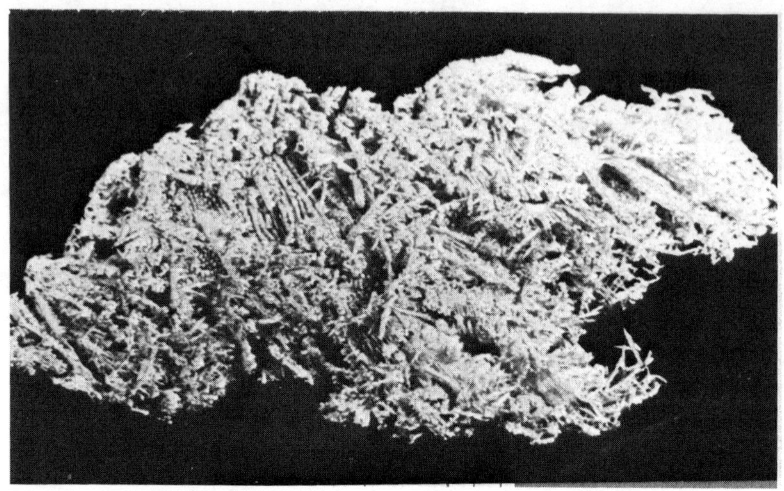

Fig. 67   *Gold, Swauk Creek mining district. Specimen is 3.5 inches long. Ernest Butler Collection.*

ILMENITE

The iron titanium oxide mineral, ilmenite, is common in metamorphic and igneous rocks as veins, masses and small grains. Crystals are thick and tabular and iron black in color. It is often magnetic and thus may be confused with magnetite with which it commonly occurs.

Subhedral to anhedral crystals to 3.5 inches long occur in pegmatite on the southeast side of Silver Creek, north of the Mineral City townsite in Snohomish County. The pegmatite is exposed in a bare rock area just above the timberline near Molybdenum Gulch. It is located on strike with the prominent ultramafic dike which is exposed to the north across the valley.

Composition of the pegmatite is biotite, quartz and orthoclase, with minor ilmenite and schorl. The schorl oc-

124

curs as small black radial masses. The ilmenite is frozen in the pegmatite and is scarce.

Rough tabular crystals occur in quartz segregations in the granites of the Golden Horn batholith along the North Cascades Highway in the Washington Pass area. The highway cuts through the batholith for several miles both east and west of the pass. The crystals occur to several inches long.

JAMESONITE

The lead iron antimony sulfide mineral, jamesonite, is commonly called feather ore in reference to its featherlike habit. The feathery appearance is formed by radiating capillary crystals; it is also found in acicular crystals or in compact masses. Its metallic luster and steel gray color may cause it to be mistaken for stibnite. Jamesonite has no cleavage, stibnite has excellent cleavage in one direction.

Jamesonite occurs in an environment similar to that of stibnite. Mineral associates are usually galena, stibnite and sphalerite. It can be found as a minor constituent in mines where lead is found.

One occurrence of good crystals of this mineral has been reported in this state. This is at the Mystery Mine in the Monte Cristo Mining District northeast of Everett. The mine is reached by taking a trail up the south bank of Glacier Creek for .75 of a mile to Glacier Creek Falls. Straight south from the falls the three dumps of the Mystery Mines can be seen 600 feet above the trail.

Outcrops of quartz diorite are abundant and make up the country rock around the mine. Galena, pyrite, pyrrhotite, sphalerite, arsenopyrite and chalcopyrite can be found as vein material on the dumps. Some of the sulfides and diorite contain small quartz veins. Much of this quartz is vuggy; the vugs are often filled with acicular crystals of jamesonite.

Mineralization in the other mines of this district is similar to that of the Mystery Mine; there may be similar specimens of jamesonite on the dumps of some of the other mines. Search only the dumps of these mines; most of them are quite old, they were abandoned in the 1930's. The timbers are rotten, and there are flooded shafts in some of the tunnels.

Break open pieces of the quartz on the dumps in search of vugs. Tourmaline crystals may be found with the jamesonite and other minerals. The tourmaline is schorl; crystals are in small radiating groups.

This is high mountain country, the Mystery Mine is located at an elevation of 4,200 feet at the bottom of a steep cliff. Wear good hiking boots and bring a camera as well as collecting tools. If the winter snowfall is heavy the road into Monte Cristo will not be open until mid-June or even mid-July. Snow is likely to cover the mine dumps until at least late July.

This is a very interesting area to visit. Monte Cristo was a town of several thousand people at the turn of the century. The owners of Monte Cristo Park are attempting to preserve this history and maintain the old buildings and ruins. It is a very scenic area to camp in and collect minerals. See update.

MALACHITE AND AZURITE

The two copper carbonates, malachite and azurite, commonly occur together. Both minerals are monoclinic and have prismatic crystals. Malachite is rare as large crystals, it is usually found as botryoidal masses. it does though, form velvet-like masses of radiating crystals. Azurite is common as botryoidal masses and rare as prismatic crystals. The most obvious physical difference

Fig. 68 *The lodge, one of the two remaining buildings from the 1880's, burned in 1983.*

between the two minerals is their colors: malachite is green and azurite blue.

Both minerals are common in the oxidized portions of copper bearing mineral deposits. Botryoidal crusts of the two minerals are quite common. The crusts rarely are coated with crystals of the two minerals. The crystals make very showy specimens, as do polished slabs of the botryoidal crusts.

These secondary copper minerals are quite common in this state, but good crystals are scarce. Good crystals of azurite have been reported, the velvet malachite can be found at a few locations. The botryoidal crusts of the minerals can be found on mine dumps throughout the state, mostly in thin layers and small pieces.

The Cardinal Mine 12 miles west of Valley in Stevens County is a source of specimens of finely crystallized malachite. These velvety coatings are common with botryoidal malachite and azurite in an oxidized zone alongside of a barite vein. Velvet malachite lines vugs in limonite. Small thumbnail specimens may be recovered at this locality.

A few small crystal groups have been found at the Deer Trail Mine near Chewelah. The crystals occur as small tufts or coatings in a vuggy limonite gossan. The crystals are scarce, but they make good thumbnail and micro-mount specimens.

A few specimens of velvet and glassy botryoidal malachite have been found at the Denny Mountain deposit. Here malachite occurs with quartz, magnetite, pyrite and limonite. This deposit has produced only a few good specimens.

At the Loon Lake Copper Mine near Loon Lake in Stevens County, velvety balls of malachite to .45 inch diameter occur with azurite and barite. Most balls have cuprite cores and form fine specimens on colorless blades of barite. The azurite occurs as bright blue crusts and as crystals to 1 inch long.

MICA

The mica group of minerals contains several species, of which muscovite and biotite are the most common. These two minerals crystallize in the monoclinic system and display perfect cleavage in one direction. God crystals of the micas are scarce; most often these minerals are found in foliated masses. Chlorite is about as abundant as muscovite and biotite, but is generally finer-grained. It is common as a secondary mineral in igneous and metamorphic rocks, as are the other two micas, and as small inclusions in quartz crystals. Lepidolite, the lithium mica, is found in lithium bearing pegmatites, mostly as scaly masses.

Muscovite has been found in groups of small crystals in the Crystal prospect near Springdale. It occurs with orthoclase, smoky quartz, and rare beryl in a pegmatite. Most of the small books are less than .25 inch across, but larger ones do occur. The specimens of quartz, orthoclase, and muscovite make fine specimens.

Good specimens of plumose muscovite can be found at the Silver Plume pegmatite at the north end of Skookum Lake east of Usk. The mine can be reached via the Boswell-King Mountain Road. Muscovite is found in a zoned pegmatite alongside a forest service road. The country rock is a coarse facies of the Kaniksu batholith.

Muscovite occurs in a plumose habit in a medium-grained graphic granite. Fine specimens of muscovite plumes were recovered when the pegmatite was mined (Fig. 69). The feathers reach lengths in excess of 6 inches and were common around 2 to 4 inches. Numerous specimens of muscovite, smoky quartz and beryl were recovered from the pegmatite.

No mining of the deposit has been done in recent years. The small pit in the pegmatite has been filled by request of the Forest Service. Very little material can be found

Fig. 69   *Plumrose muscovite in graphic granite, Silver Plume Mine, Usk. Specimen is 3.5 inches high.*

around the pegmatite. If reopened in the future, some fine specimens may again reach the market.

An occurrence of lepidolite has been reported at LaBohn Gap at the head of the Middle Fork of the Snoqualmie River. The authenticity of this report is not known. A deposit of chalcopyrite is being prospected at this locality; Schorl has been reported, too. There may be an interesting pegmatite here.

MIMETITE

Mimetite is an uncommon mineral which occurs in oxidized portions of lead-bearing ore veins. usually it occurs as radiating clusters of acicular crystals. Sometimes it forms mammillary crusts or rounded crystals known as campylit. The color is usually white to yellow, orange or brown.

Fig. 70 *Mimetite, Cleveland Mine. Center cluster is .4 inch across. Rudy Tschernich Collection.*

Fine clusters of micro-sized crystals can be found at the Cleveland Mine near Hunters, (Fib. 71b). The centers of the clusters are light brown; the edges are light yellow, (Fig. 70). Other minerals found here include cerussite, austinite, and hemimorphite.

MOLYBDENITE

Molybdenum sulfide, molybdenite, is common in many copper mineralized areas. It is found associated with chalcopyrite, pyrite and quartz. It is common as an accessory mineral in granitic rocks and pegmatites. The metallic luster, lead-gray color and perfect micaceous cleavage make it easy to recognize. Crystals are hexagonal shaped plates or short tapering prisms. Most commonly it is found as foliated masses or scales.

131

Fine specimens of molybdenite on diorite with quartz have been found in the Crown Point Mine at Holden. When in production this mine produced numerous specimens with hexagonal plates up to 1.25 inches across. Quartz crystals occurred with the molybdenite on matrix. The mine is closed but there have been a few specimens collected in recent years.

At the Weden creek-Silver Creek divide and south of there at the end of the logging road on Silver Creek, molybdenite can be found in joints in diorite of the Snoqualmie batholith. The locality at the divide produces a few quartz crystals. At both localities the molybdenite occurs as plates that either coat the joint surface or stand on edge on the surface. The plates are less than 0.45 inch across. Some large specimens may be found coated with molybdenite.

This is an interesting area with several small mines east of Index, on the Garland Mineral Springs road, the highway cuts through a molybdenum occurrence that contains numerous vugs with small quartz crystals and molbydenite crystals. The deposit is located just before two bridges. The highway takes a sharp right turn around a small cliff. There is a large pile of diorite boulders on the left side of the road.

Crystals can be found in the vugs in the boulders and in joints in the cliff. Molybdenite, up to .75 inch across, small quartz crystals and massive chalcopyrite may be found at this locality. The molybdenite plates may be partially intergrown with the quartz, so that small grooves remain in the quartz if the molybdenite is removed.

Collectors will need a hammer and chisel to collect these specimens. The surface material in the boulders has been heavily picked over, by breaking the boulders, new material will be exposed. Few collectors have searched the joints in the cliff for specimens, which are difficult to collect because of the bulk of the rock. Individual crystal plates can be removed, but it is difficult to remove matrix specimens.

PALYGORSKITE

Palygorskite, a very unusual mineral, is often called "mountain leather". This complex hydrous magnesium aluminum silicate occurs in very flexible fibrous sheets. Some of these are very smooth and soft, similar to doeskin; thicker sheets are stiffer and rough. It is commonly light to dark gray in color.

In Washington, palygorskite is found in the Pend Oreille Mine and the Josephine Mine near Metaline Falls. Here the mineral forms thin sheets in fractures in the limestone of the mine. In the lower level of the mine it also is found with calcite and quartz crystals in vugs (Fig. 62). Many of the sheets that can be recovered are over 1 inch in thickness and nearly 3 feet square. Some specimens from the vugs may be found with calcite crystals, small quartz crystals or pieces of black dolomite attached to the palygorskite.

Most the of calcite crystals are less than .25 inch across, but larger ones have been found associated with the palygorskite. The quartz crystals are quite small and stubby, less than .75 inch long and milky. Small sphalerite crystals, almost black in color, have been found on some specimens. Most of the palygorskite is not associated with good crystals of these other minerals.

PYRITE

The most common sulfide mineral is probably pyrite. It is common in ore deposits associated with quartz and in metamorphic and sedimentary rocks. This iron sulfide crystallizes in the isometric system. Crystals are usually cubes, octahedrons or pyrithohedrons that are often intergrown or twinned. Sometimes pyrite occurs in granular,

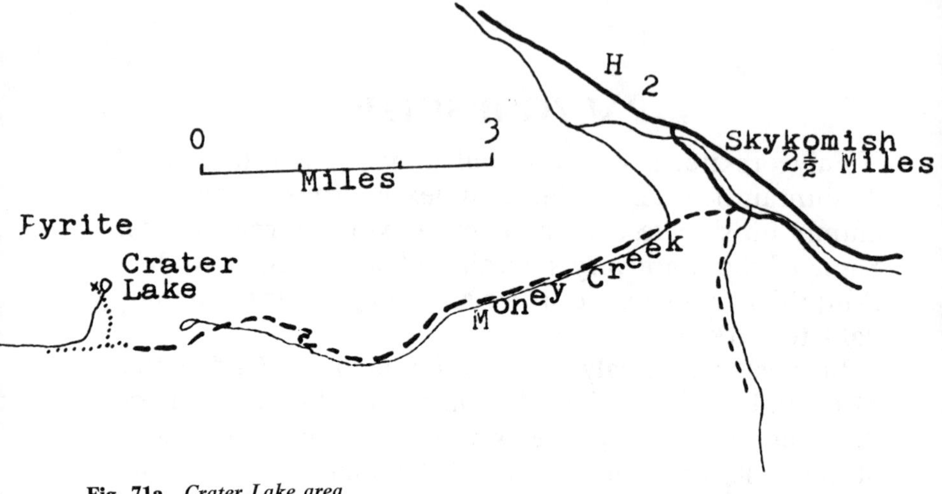

Fig. 71a *Crater Lake area*

Fig. 71b *Hunters area*

reniform or stalactic forms that make excellent specimens.

All types of rocks form the hosts of pyrite which may be a primary or secondary mineral. It is found finely disseminated in rocks, in vugs, in coarse granular lenses or as large crystals in rocks or vugs. Crystals are most common up to .5 inch on an edge, but are often much larger.

In Washington, pyrite is found associated with other minerals in several of the mines. Few mines have reported notable crystals, but there are numerous mine dumps that may be searched by the collector for an occasional fine specimen.

Undoubtedly the finest Washington pyrites are found with quartz crystals at Spruce Ridge on the Middle Fork of the Snoqualmie River. This locality is described under "Quartz Localities". The pyrite occurs as cubes and pyritohedrons up to 6 inches across. Many museum quality specimens in clusters of quartz crystals have been collected here.

The deposit is a breccia exposed in a steep fault canyon. Vugs up to 4 feet in diameter and 20 feet long often produce hundreds of fine specimens. The most common crystal habit of Spruce pyrite is cubes modified by pyritohedrons, with the two forms being equally developed. Highly striated cubes, and pyritohedrons with minor octohedral and trisoctohedral modifications also occur.

At Denny Mountain a few pyrite crystals have been collected with quartz and other minerals in a skarn zone. The pyrite varies from fresh to completely altered to goethite. Specimens can be found in vugs that are filled with calcite and in clusters with quartz crystals. These crystals are usually less than .25 inch across. A more thorough description of the deposit is given in the quartz chapter.

East of the Tolt-Seattle Reservoir in King County on the west side of Crater Lake, there is a large breccia zone in volcanic rocks that contains pyrite (Fig. 71a). Few crystals can be found in vugs in the cliff. The crystals are less than .25 inch across, but they occur in clusters that make fine specimens. The mine is flooded with about 18

Fig. 71 *Pyrite, Spruce Ridge, 3.4 inches across. Bob Jackson collection.*

Fig. 72 *Quartz and pyrite, Suiattle River. Pyrite cube is 4 inches across.*

inches of water. Collecting may be difficult and specimens scarce.

Small crystals can be found in vesicles in the Columbia River basalts in eastern Washington. Some fine micromount crystals can be found in the basalt quarry at Granite Point upstream from Wawawai on the Snake River. The crystals are in the middle layer of massive basalt in the quarry above the columnar layer. Crystals are less than .1 inch on a side, most are cubes. The cubes are usually intergrown in groups resembling roses or building blocks. Most of the cubes do not show any striations and are very bright and shiny. The basalt is very hard, so heavy tools will be needed to collect these specimens. The basalts in this locality may be prospected for vesicles that contain other minerals of interest.

Interesting filamentous pyrite can be found with paulingite and other minerals at the Rock Island Dam near Wenatchee. See the zeolite chapter for a more complete description of this locality. The pyrite occurs as filaments in vesicles. Some of the filaments make a right angle bend, but most crystals are straight and only .1 inch long. They make interesting micromount specimens.

Large pyrite crystals have been reported from a quartz deposit on the Suiattle River (Fig. 72). The deposit is reportedly more than half way up the river in a very difficult to reach area. The cubes are partially altered so that they have irregular faces. They are unusual in the fact that they are shiny and not coated with iron oxides. The alteration process is removing the iron and leaving elemental sulphur on the pyrite faces.

Fine specimens of pyrite can be found in altered tuff along McCoy Creek southwest of Randle. The location is accessible via a Forest Service Road, Road 1013, and a dirt road. The pyrite occurs as singles and clusters in clay seams. Specimens reach several inches on an edge and are very fresh and shiny. Most specimens have deep striations and a majority are cubes. A very few are truncated by octahedrons. They can be collected by digging through several feet of soil along the stream bank to the tuff. Good digging tools are needed and collectors must be aware of the danger of holes caving in. Hard work at this locality will reward the collector with some excellent specimens.

PYRRHOTITE

Pyrrhotite is common in metallic mineral deposits. Most often it occurs as irregular masses, rarely as euhedral crystals. It crystallizes in the hexagonal system with the crystals usually thin hexagonal plates, large crystals are quite rare. It has an iron sulfide composition, is bronze in color and magnetic. The magnetism and low hardness, 4, aid in distinguishing it from pyrite.

Although pyrrhotite can be found in many of the mines of the state, crystals have not been reported in any of the mining districts. The only occurrence of crystals are in vesicles in the basalt around Spokane and possibly other locations in eastern Washington. In Spokane it occurs with siderite, cristobalite, feldspars and opal. In the vesicles it forms small hexagonal plates that are less than 1 mm. in diameter. It is not rare, so a few specimens can be found in at least one of any group of vesicles.

These crystals make good micromount specimens. Collecting locations and required tools are described under siderite. Similar crystals may occur around Pullman, Pasco and other locations in the Columbia River basalts of eastern Washington.

Excellent hexagonal crystals to .75 inch diameter occur at a small prospect 13 miles northeast of Index next to the Silver Creek Road. Fine specimens displaying clusters of crystals occur here. Other mines along Silver Creek produce sphalerite, arsenopyrite, galena, pyrite and quartz.

QUARTZ

Quartz has a wide appeal as a mineral specimen as well as a gem. Crystals of quartz occur individually and as intergrown groups. The most common habit of quartz is the hexagonal prism terminated by a combination of positive and negative rhombohedrons. This form is often distorted, as in tabular crystals, which have two parallel hexagonal prisms greatly overdeveloped as compared to the other four. Japan Law twins of quartz often show the tabular habit. In this type of twin the c-axes are inclined at 84°33' (Fig. 73).

Quartz scepters are very desirable as specimens. These are crystals which have a section of prism and termination of a larger diameter than the main prisms (Fig. 74). This is usually caused by two or more separate episodes of crystal development. The smaller diameter crystals may be visible inside the scepter termination. Often the shaft and termination may be different colors; a rock crystal shaft with an amethyst termination is a common form.

Phantom crystals are another form of quartz much sought after by collectors. In this form the crystals have the usual hexagonal prism with the outline of another termination visible inside of it. This outline reveals the presence of an earlier crystal which was completely overgrown by a later generation of quartz. The outline or phantom may be emphasized by fluid inclusions, inclusions of another mineral, or color variations within the crystal. A close look at most amethyst crystals will reveal numerous phantoms (Fig. 65b).

Crystalline quartz is given several varietal names dependent upon its color. The mauve to purple variety is known as amethyst; the brown to black as smoky quartz or morion. (The jewelers' term "smoky topaz" refers to smoky quartz). Colorless quartz is known as rock crystal; cloudy-white material is called milky quartz. All of the above varieties are found in Washington.

Other quartz crystals of particular interest to collectors are those that contain inclusions of other minerals. Common inclusions are chlorite, pyrite, and iron oxides. Several locations for these and the above mentioned varieties are described below.

Fig. 73 *Quartz, Japan Law twin, on prismatic quartz, Denny Mountain. The tabular twin is 5mm high.*

Western Washington
Middle Fork Of The Snoqualmie River

The largest quartz-producing area in the state is in breccias along the Middle Fork of the Snoqualmie River, about 45 miles east of Seattle. Most of this steep mountainous region is included in the Alpine Lakes Wilderness area, where collecting is controlled by the US Forest Service. Collectors must contact the North Bend Ranger District (phone # 888-1421) BEFORE venturing into the wilderness to collect. Most of the breccias are covered by mining claims.

Fig. 74 *Quartz "Raspberry" scepter, Denny Mountain. Note needle like inclusions and long fluid inclusion.*

One of the two most productive deposits is a large breccia near Green Ridge. This is a deposit of diorite breccia, part of the Tertiary age Snoqualmie Batholith. There are actually two breccia bodies separated by a small ravine. The largest breccia forms several cliffs and talus slopes (Fig. 76). Both breccias are composed of weathered fragments of a light gray diorite. Vugs are abundant and some are quite large. The crystals which line the vugs

Fig. 75   *Middle Fork of the Snoqualmie River.*

average 2 inches long and .75 inch diameter. Most are clear in the top 2/3 of the crystal; some are milky and a few are snow white. Amethyst is present but uncommon.

Most of the crystals are of the common prismatic habit, but a few doubly-terminated crystals and scepters have been found (Fig. 77). A few fine groups of phantoms outlined by fluid inclusions have been discovered over the years this deposit has been worked.

The clear crystals are often of faceting quality. A friend of the author found a 3.5 inch flawless crystal in the talus, but few facet-quality crystals reach this size. Vugs of limonite and clay have been encountered which contain clear crystals, some doubly-terminated. The limonite-clay mixture is easy to remove from the vugs when it is wet, but next to impossible when it is dry.

A large vug found by the author contained amethyst crystals on the roof with rock crystal lining the sides and bottom. Most of the amethyst crystals were nearly doubly-terminated, being attached to the diorite matrix by an incomplete termination. Some reached 2 inches long by 2

142

Fig. 76 *The author on the quartz-bearing breccia, Green Ridge.*

Fig. 77 *Quartz, showing double-termination and parallel growth, Green Ridge.*

inches wide. When found they were coated by a layer of hematite and other iron oxides.

Four mining claims cover the breccias at Green Ridge. As commercial specimen-producing operations often occur here, it is imperative to contact the owners before undertaking the long hike to the area. Claim ownership is public information and can be obtained from the appropriate US Forest Service district office, or from the Bureau of Land Management in Portland. All quartz localities near Green Ridge are within the Alpine Lakes Wilderness area. Permission to collect must be secured *in advance* from the US Forest Service in North Bend.

Numerous other deposits which produce excellent quartz crystals are located along the Middle Fork. Four of the best are: Clipper, Spruce Ridge, Katie Belle, and Pedro Pipe. The Spruce Ridge deposit has been producing fine quartz and pyrite groups for many years. Since 1976 it has been worked by commercial collector Bob Jackson, of Renton, WA. Jackson provides group collecting trips during the summer; Spruce is closed to collecting except on these trips.

Quartz crystals at Spruce average 2 inches in length, and 3/8 inch diameter. Most have a tapered habit. Scepters are common, and Japan Law twins have been found. Doubly terminated crystals are very common, as are crystals with pyrite inclusions. Accessory minerals are calcite, ankerite, and barite. It is possible to collect plates showing all five minerals on a single specimen.

The average vug is about 6 X 6 X 6 inches, but tubes up to 4 feet in diameter and 20 feet long have been found. Vugs large enough to crawl into are a popular feature on Jackson's trips.

A second quartz breccia exposed at Spruce Ridge produces thousands of lusterous quartz crystals to 1 inch, over 50% of which have pyrite inclusions. The crystalline pyrite apparently coated terminations before the solutions which deposited quartz were depleted. Many of these small crystals are doubly-terminated, and highly sought after by jewelers.

The finest quartz and pyrite specimens from Spruce Ridge grace museum collections world wide. Although quartz and pyrite are both common minerals, they seldom occur together in well crystallized form. The best collection of Spruce material on public display in the Northwest is at the M.Y. Williams Geologic Museum (University of British Columbia), Vancouver, B.C., Canada.

About 1.5 miles east of Spruce Ridge there is a deposit of quartz on a patented copper claim known as the Clipper. At this point several logging roads strike off north from the Middle Fork road through a logged off area. Just above the uppermost of these roads is a small adit, which marks the southernmost extremity of the Clipper quartz deposit. Most of the crystals often act as supports for attractive small groups of geothite or pyrite. The best collecting area is found on a sharp prominence about 100 feet above the adit, among large trees.

Just east of Hardscrabble Creek, about one mile east of Clipper, a small breccia pipe is exposed on the bare sides of a steep ridge. This is the Pedro Pipe. Access to this deposit is very difficult due to the steepness of the terrain. Quartz crystals similar to those of Clipper are found here, together with fine specimens of pyrite, molybdenite, and galena. Unless the collector is familiar with rock climbing techniques and equipment, this area is probably too dangerous to attempt. (Fig. 78 is of the author's camp above the pipe; vertical cliffs surround the tent on three sides).

Fig. 78 The author's camp on the south side of Katy Bell Ridge. The flat spot where the tent is pitched is an old helipad.

147

Katie Belle Ridge has several small breccias exposed along its top and south side. All of these areas, and the Pedro and Clipper localities, are on private land. The US Forest Service North Bend office can provide an up-to-date list of owners names, as well as grant permission to collect on non-claimed areas within the Alpine Lakes Wilderness. All of the localities in the Middle Fork area are either held under mining claim, or within the boundaries of the Wilderness area. Several collectors have been fined for collecting without permits: contact the US Forest Service BEFORE you collect!

There are at least two quartz crystal localities in the Myrtle Lake area, north of Mt. Price. This area is reached via the Dingford Creek trail. The trailhead is on the Middle Fork road about six miles east of the Taylor River campground. A hike of about 3.5 hours is required to reach Myrtle Lake, and an additional hour to reach Little Myrtle Lake. Both lakes offer good camping spots and good fishing. The trail is in good condition, being well-used by backpackers and fishermen. From Little Myrtle Lake an hour's hike is necessary to reach the quartz locality west of the lake. The crystals are found in two spires on the ridge crest. The tops of these are visible from the mountainside about 50 feet above the east shore of Little Myrtle Lake. Take this little climb to determine the best route to the spires. There is no trail, the route is steep but not difficult. Head directly for the spires until a talus slope is reached, then skirt around the southeast edge of the talus and follow it up to the spires. The best collecting is in the talus boulders on the west side of the spires (Fig. 79). The crystals are weathering out of near horizontal quartz veins which cut the spires.

Good undamaged crystals can be worked out of the veins or the talus boulders. Most are under one inch in length and .5 inch diameter, associated with pyrite and limonite. A few crystals are doubly terminated. Large flat plates can be collected here. This deposit has received little attention from collectors.

There are scattered quartz veins along the ridge north

Fig. 79 *Spires on the ridge west of Little Myrtle Lake.*

Fig. 80   *View east from Big Snow Mountain, east of Myrtle Lake.*

Fig. 80a   *Ridge atop Denny Mountain, foreground, as seen from Granite Mountain
(Mike Miller photo)*

Fig. 80b   *A collector in an empty vug. Denny Mountain. Note bases of quartz crystals
on the right side of vug. This vug is about 50 feet above the canyon floor.*

of the spires. A few pegmatitic dikes occur on the east bank of the creek which connects Myrtle and Little Myrtle Lakes and extend some distance to the east. Crystals are abundant in the talus and cliffs. Large blocks of limonite which appear to be pseudomorphic after pyrite occur also. No fresh pyrite was visible, but it might be present at depth.

The view from the ridge and mountain tops in these areas is magnificent. Numerous lakes and peaks can be seen from these vantage points (Fig. 80). The quartz deposits are located at elevations from 3,000 to 4,500 feet. Winter shows reach depths in excess of 25 feet in some places. For this reason, most of this area is accessible only from July to mid-October.

South Fork of the Snoqualmie River

There are at least two good quartz producing localities along the South Fork of the Snoqualmie River. Denny Mountain is the best known deposit; it has produced quartz crystals and other minerals since the turn of the century. The deposit was considered for production of quartz crystals for radio use during the Second World War, but its size proved too small and there is no record of any production. (Fig. 80a).

The collecting area is a long, steep, deep, narrow canyon which exposes a large skarn zone at the contact of Snoqualmie granodiorite and a limestone body. The area has undergone several episodes of hydrothermal alteration and has produced grossular, epidote, pyrite, clinozoisite, diopside, scheelite, hematite, goethite, malachite, chrysocolla, chalcopyrite, bornite, albite and several interesting varieties of quartz. The canyon has been nicknamed "Rockhound Gulch" by the Forest Service, in recognition of the fact that ten collectors have lost their lives in this canyon during the past 20 years. The productive areas are covered by several patented mineral claims. This area has recently been added to the Alpine Lakes Wilderness area. Contact the US Forest Service North Bend office for collecting regulations.

There are three separate mineral-producing areas within the canyon. They are described here beginning with the lowest in the canyon, that is, the first a collector walking up the canyon would encounter.

Near its confluence with Denny Creek, the canyon is quite wide and brushy. It narrows just below an outcrop of massive magnetite, just above a waterfall which the collector must climb around. This area is commonly called the Red Wall, and contains vugs and seams of grossular garnet and pyrite. The garnets are orange to brown in color, and average .25 inch in diameter. They mostly exhibit trapezohedron and dodecohedron faces, but cubes have been reported. Some large masses of garnet contain vugs with epidote or clinozoisite blades. The blades are dark green in color and often reach three inches. A few are terminated, but most extend completely across the vugs. The vugs also often contain small milky quartz crystals and adularia. Most are completely filled with calcite, which may be disolved away with hydrochloric (muriatic) acid to reveal the other minerals. Most of the calcite fluoresces light pink under short wave ultraviolet light. A few vugs have yielded pale yellow sheelite bipyramids to .25 inch and green diopside crystals to .75 inch.

The upper end of this zone contains numerous quartz veins exposed high on the cliff. Vugs over 3 feet in diameter are related to this vein, and contain quartz, pyrite, and goethite psuedomorphs after pyrite. The quartz crystals commonly taper to their terminations and are gray or etched on the outside, but transparent inside. Some are over a foot long and 3 inches in diameter. Pyrite cubes over 2 inches on a face have been found associated with the large quartz. Most of these vugs have been cleaned out years ago; frost action removed the large crystals from their matrix, so few groups of these large crystals are in existence. Working in the bottom of one of these old vugs, the author and a friend removed over 50 pounds of large terminated crystals in 1975 (Fig. 81). A constant trickle of water runs out of these vugs, which makes access from the canyon floor very difficult due to a growth of slippery moss on the near vertical canyon sides. (Fig. 80b).

Just to the left (east) of these vugs a small stream enters the main canyon from a narrow and steep canyon. Up this canyon are additional vugs of quartz and goethite pseudomorphs after pyrite, but the vugs and crystals are considerably smaller. Several years ago, a single vug filled almost entirely with spectacular Japan Law twin quartz crystals was found (Fig. 73), but subsequent work in the area has revealed no additional twins.

At the confluence of the small canyon and the main canyon, the main canyon turns left (northeast) and the collector must climb above a steep slippery waterfall. It is

Fig. 81 *Quartz and pyrite, Denny Mountain. The longest crystal is 7.25 inches long.*

highly inadvisable to continue above this point unless you have climbing equipment and are familiar with its use. The talus slope above the waterfall is extremely unstable; several deaths have occurred at this point, collectors being crushed by rolling boulders.

About 200 feet further up the main canyon is one of the most interesting occurrences of quartz in the state. Amethyst quartz scepters occur in a zone of specular hematite with calcite, epidote and grossular. The grossular and epidote are of the same habit as that from the Red Wall below. Calcite fills vugs and can be dissolved away to reveal quartz and other minerals. Hematite occurs in large micaceous "roses" up to three inches and as massive veins and cavity fillings.

The quartz scepters lined a continuous series of vugs atop a hematite vein. Most were white, but some spectacular amethyst scepters have been found. These ranged from light purple to bright red, and contained sparkling inclusions of hematite and lepidochrocite, causing them to be given the local name of "Raspberry Scepters". These vugs were mined out in the late 1960's, and the chances of finding spectacular specimens today is very slim. A huge amount of rock would need to be removed before any additional vugs could be discovered, if indeed any additional vugs exist. Occasionally a small single scepter is found in the talus below these workings (Cover photo). See update.

Apparently there were at least three generations of quartz deposited in this zone. The first quartz crystals deposited were of the normal hexagonal prism habit. This quartz was clear to slightly milky. The second generation was milky to gray and overgrew the first on four or five prism faces and totally coated the terminations. This produced crystals which have a "hooded" appearance when viewed from the uncoated prism faces (Fig. 82). This type of crystal is very abundant in the talus of the scepter area as they are considered waste when encountered during the search for "raspberry" scepters. These white crystals are locally termed "overgrowth scepters." A third generation of quartz deposited the amethyst scepter terminations, as well as the inclusions of hematite and lepidochrosite. These characteristic inclusions serve to positively identify the Denny Mountain scepters from all other amethyst scepters. The hematite inclusions usually consist of silver needles of irregular outline, the lepidochrosite are flattened plates of which one end is usually delta or "Y" shaped. A cross-section of any of the "raspberries" clearly indicates the multiple generations of quartz.

Denny Mountain is one of the most interesting deposits in the state, and undoubtedly the most deadly, though not necessarily the most dangerous. Its easy accessability from the Melakwa Lakes Trail makes it attractive to collectors who don't have the experience to deal with its dangers. The collector who visits Denny Mountain should be familiar with rock climbing, should wear a hard hat and sturdy boots, and should use EXTREME CAUTION when moving any rock on a talus slope. The author has often

Fig. 82 *Quartz scepters, Denny Mountain. Largest crystal 2.4 inches long.*

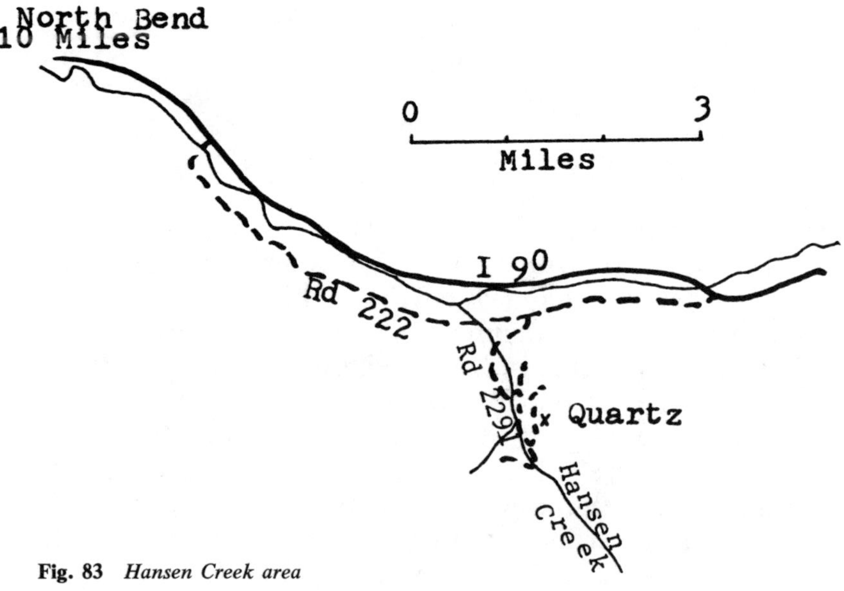

Fig. 83 *Hansen Creek area*

witnessed rock falls from the high cliffs; collectors must be wary of these ever present dangers.

Another locality which has produced excellent single amethyst scepters is Hansen Creek, east of North Bend. The scepters from this locality are highly skeletalized and misshapen. Most are of a pale amethyst color, which highlights distinct zoning. Also found are doubly terminated quartz crystals and crystals with chlorite and/or pyrite inclusions.

This area is believed to be a large slump zone which has moved down the mountain. The area is still quite unstable, as can be seen by the development of large cracks and slides in the logging roads. Crystals are found loose in the clay and sandy soil of a logging road cut. Boulders of diorite which contain quartz vugs are numerous. The diorite is the likely source of the amethyst scepters and the other crystals, but the author is not aware of any large crystals having been found in place in the diorite.

This deposit is reached by taking Interstate 90 for 11.5 miles east of North Bend to where a road turns south. This road, the Tinkham road, joins another which accesses the Hansen Creek road in about 5 miles. Two localities exist at Hansen Creek: one a group of mining claims on Forest Service land, and another on Weyerhauser land. Damage to a forest road has caused Weyerhauser to close its locality; access to the mining claims are by fee arrangement. Owners can be contacted through the US Forest Service North Bend office.

Two other quartz localities are known in this area, though they are seldom visited due to difficult access. Both are north of Denny Mountain, and may be reached by continuing on the Melakwa Lakes trail through Melakwa Pass. Guye Peak, located in T23N, R11E, contains a skarn zone similar to that of Denny Mountain, but without the variety of material present. Quartz crystals up to six inches in length have been found in this skarn, but to the author's knowledge, no scepters or twins have ever been recovered. The other locality, Mt. Roosevelt, located in T23N, R10E, produces quartz, epidote and stilbite from a shear zone near the peak. Crystals are not abundant at this locality, and rock climbing gear is needed to reach it.

North Fork of the Snoqualmie River

Several good quartz crystal localities exist along the North Fork of the Snoqualmie. Mt. Teneriffe, northeast of North Bend, produces a good quantity of Japan Law twins, some of which are quite large. Normal habit crystals, some with tourmaline inclusions, are also abundant. The deposit is located on the northeast side of Mt. Teneriffe on a very steep slope. A .5 mile hike is required to reach this breccia zone. Access from North Bend is by taking the North Fork road to Weyerhauser gate 10. Turn right and cross the bridge to road 4200. Follow 4200 to 4204 (right turn) and 4204 to 4220. Park at the end of 4220 and hike southwest through the clearcut up the hill to the edge of the wooded area, then traverse the hillside heading west at the edge of the wooded area until a talus slope is

Fig. 84   *Quartz cluster, Mt. Teneriffe. Note Japan Law twin in center of cluster. Twin measures 1.4 inches high by 2 inches wide. Rudy Tschernich Collection.*

reached. At this point it will be necessary to hang onto brush and trees to traverse the talus slope to the crystal-bearing outcrop. A similar area is located on the ridge line between this point and Haystack Mountain to the west; it may be reached via logging roads from the Sallal Prairie road.

The Mt. Teneriffe locality is located on state land. Permission to collect must be secured in advance from the Dept. of Natural Resources, Olympia. Weyerhauser Company controls the access roads. No camping is permitted.

These are hard rock deposits requiring much work to produce significant specimens. The area is quite popular with collectors, and should continue producing for many years. Crystals average 1.5 inches long by .5 inch in diameter. Clusters of crystals with attached twins 2 or more inches high have been collected by the author. (Fig. 84).

At the end of the North Fork road is Devil's Canyon, a locality which has produced Japan Law twins and normal quartz crystals. The area is claimed as a copper-molybdenum property; the claim owners have posted their property and do not welcome collectors. A very steep narrow canyon exposes numerous quartz veins in diorite of the Snoqualmie batholith. Vugs are quite common in some areas; a few contain pyrite and molybdenite crystals along with the quartz. The largest crystals have been found in the canyon, however, vugs are present on some of the ridges in the vicinity of the canyon.

A breccia zone in diorite on Prospectors ridge north of Lennox Creek contains quartz crystals. The crystals are clear and up to 1.75 inches long. Chlorite and epidote are associated.

In Illinois basin, quartz occurs with epidote in miarolitic cavities. These occur in the southern and western parts of the basin. Abundant outcrops and talus slopes provide good collecting.

Other Areas

Quartz crystals have been found in most of the mining districts of western Washington. Three districts of current importance are: Index, Sultan Basin, and Monte Cristo. The area around Index, Snohomish County, is peppered with old mine workings and prospects, many of which are visible from the roads paralleling the Silver Creek drainage. Quartz was an important gangue mineral in this area, and occurs on many of these tailings piles. Of particular interest are those workings near the head of Silver Creek, in an area called Red Gulch. The collector who climbs to the head of Red Gulch will encounter several quartz crystal producing areas. The gulch ends at an abrupt ridge. The view from this ridge is spectacular. Mt. Baker can be seen to the north and Mt. Rainier can be seen to the south. There are often mountain goats in this area, they can be seen sunning themselves in the rocks. From this area collectors can see into the nearby mining districts of Monte Cristo and the Sultan Basin.·

Large crystals have been found in the Bren Mac Mine in the Sultan Basin. It is presently being explored for the production of copper. The mine is being developed in the

Sunrise Breccia Pipe, which is in the Vesper Peak stock, a satelite of the Snoqualmie batholith. The breccia is composed of angular fragments of hornfelsed metasediments of the Chilliwack group.

Fragments of the breccia are cemented with quartz, chalcopyrite, molybdenite, scheelite and pyrrhotite. Vugs are numerous in parts of the stock. Crystals of quartz line most of the vugs, and a few small dolomite crystals have been found in others. In one spot in the breccia mine, the author noted two clear quartz crystals about 2.5 inches in diameter embedded in massive milky quartz. The crystals were exposed in cross section in the wall of the tunnel, their length is unknown.

During driving of the main mile long tunnel several crystals were recovered. A fine group of these crystals is on display in the Department of Geology and Earth Resources offices in Olympia.

Small amethyst crystals can be found on Dog Mountain, Skamania County. The crystals are found in small radiating clusters of light amethyst color. Most of them are less than 1 inch long and .25 inch in diameter, and taper from base to termination.

Some excellent barite, quartz, and zeolite specimens have been recovered from volcanic rocks at Kosmos. A large pocket of quartz pseudomorphs after calcite crystals was removed from a river bank near Kosmos several years ago. A thorough search of the area has revealed no more of this material.

A few quartz crystals have been found on Yellow Aster Butte in Whatcom County. Crystals up to 4 inches long have been recovered from a quartz vein in metamorphic rock. The location is near the south ¼ corner of section 18, T40N, R9E. (Huntting, 1960)

Good quartz crystal geodes, some amethystine, have been found at Walker Valley, near Big Lake. This locality is described in the cryptocrystalline quartz section of this book. The geodes reach a diameter of 12 inches. Most of the quartz crystals are clear, but light to dark amethyst is also common. (Fig. 85).

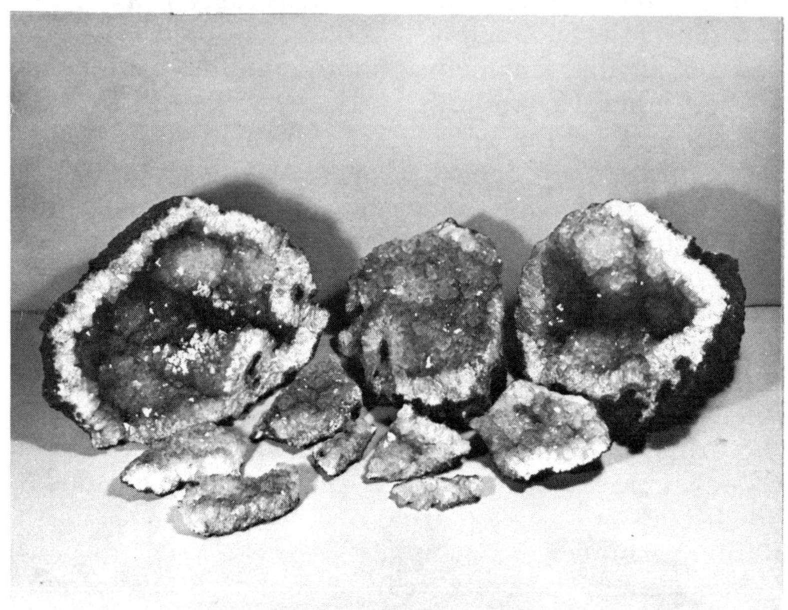

Fig. 85 *Amethyst geode sections, Walker Valley. Left specimen is 14 inches in diameter. Mike and Gloria Johnson Collection.*

Fig. 86 *Goethite on quartz, Walker Valley. Goethite tufts are 2mm high.*

Fig. 87 *Quartz rods in geode, Walker Valley. Field of view is 1.5 inches. Note small calcite rhombs on the quartz.*

Fig. 88 *Smoky quartz with aquamarine, Crystal Prospect. Gus Girnus Collection.*

163

Many of the quartz crystals in the geodes are coated with small tufts of brown goethite (Fig. 86), or contain radiating tufts of black to grown goethite as inclusions. Most of the goetithe tufts are less than .25 inch long. Rhombohedral calcite crystals can also be found in some of these geodes. These crystals are clear to white or stained a light yellow with iron oxides. A few geodes are completely filled with calcite.

Some geodes are filled with stalactites or rods of quartz (Fig. 87). The rods are formed by small quartz crystals that radiate from a carbonate core. These make interesting specimens.

These geodes are found in a basalt cliff and must be removed with a large hammer, chisel and bar. The deposit is on state land and is held under a claim by the Washington State Mineral Council. All collectors who belong to a club are welcome to collect at this locality.

Two locations for quartz are reported in Jefferson County. Mount Anderson in the NW¼ SW¼ of sec. 28, T26N, R5W has some crystals under 2 inches in length. The other locality is along Rustler Creek, in the SW¼ SW¼ of sec. 31, T25N, R7W. A quartz vein here contains crystals up to 2 inches in length and 1.5 inches in diameter. These two localities are within the Olympic National Park. Collecting is not allowed. This area is being reconsidered for Wilderness classification; if it is reclassified, then collecting would be allowed.

McCall Basin north of the Goat Rocks in the Cascade Mountains has produced some small quartz crystal clusters in vesicles in a green altered volcanic rock. This area is quite large. The crystals are mostly less than .5 inch long and are a bright white color. The clusters often include four or more crystals in vesicles up to .4 inch in diameter.

Northeastern Washington

Smoky quartz crystals can be found in a pegamite on the Crystal Prospect near Springdale, 32 miles northwest of Spokane. This pegmatite is on private land, but collecting has been allowed. The small crescent-shaped dike intrudes a quartz monzonite phase of the Loon Lake Granite (Patee, Van Noy, and Weldin, 1968) in sec. 15, T26N, R38E.

Quartz crystals occur in numerous vugs in the pegmatite, which is composed of orthoclase, quartz, muscovite, and beryl (rare). The crystals are clear, gray, or smoky. One large vug was found which contained several hundred pounds of crystals up to 12 inches long, on and off matrix (Fig. 88). Two small aquamarine crystals were found associated with the smoky quartz.

Muscovite occurs as euhedral crystal books in the vugs. Many of these books are up to one inch across and .25 inch thick.

Many good specimens have been extracted from this pegmatite. Collecting is done by digging in the partially decomposed surface of the pegmatite and by breaking the fresh rock in search of seams or vugs. A few small crystals can be found by most collectors in a day.

Near Usk, 50 miles north of Spokane, are two localities which have produced large smoky quartz crystals; the Silver Plume Pegmatite and the Dutton Prospect. The Silver Plume deposit is named for the plumose muscovite that is abundant in part of the pegmatite. The deposit is near North Skookum Lake on the side of the road. It is held under mining claim.

During mining of the muscovite, a large vug was encountered which contained smoky quartz crystals. The largest crystals were a pair of parallel grown dark smoky crystals about 18 inches long and 12 inches in diameter. They were not terminated, apparently both ends had contacted the walls of the vug. Several smaller clear and smoky crystals were found.

A few beryl crystals were collected during the mining operations, along with some autunite. The beryl crystals were up to 4 inches long and green in color. They were not of gem quality.

Smoky quartz crystals to 4 inches long occur in the pegmatites at the Dutton Prospect on South Baldy Mountain. Crystals occur in soil over decomposed monzonite and pegmatite dikes. Associated minerals include beryl, chrysoberyl, orthoclase and albite. Numerous pits have been dug on the claims exposing pieces of pegmatite, but no dikes have been found in place.

Access to the locality is by taking the Kings Mountain Boswell Road from Usk for 14.3 miles, then going .5 mile north on the South Baldy Mountain Road. The pits are .5 mile northwest of this point on a road that extends up the hill.

In the Sherman Pass area, east of Republic, clear to smoky quartz crystals occur in pegmatites. The crystals reach lengths of 2.5 inches and some are of a complex skeletal form. They occur with green microcline, schorl, spessartine and pyrite.

On Topaz Mountain and Windy Peak north of Winthrop, smoky crystals occur in miarolitic cavities and vuggy aplite dikes (Cannon, 1975 and personal communication Bob Jackson, 1975). Associated minerals include albite and spessartine. Reportedly, topaz was found in the past.

REALGAR

Realgar is not a rare mineral, but it rarely occurs as euhedral crystals. It is commonly found with other arsenic minerals: orpiment and stibnite and with cinnabar in low temperature veins. Crystals are monoclinic, vertically striated, and prismatic. The color is orange to red and quite showy. The bright red crystals are usually not of large size, being mostly under .25 inch in length. Unfortunately realgar is unstable, especially in the presence of light; crystals alter to yellow orpiment and crumble apart. For this reason realgar does not make good specimens for permanent display. Realgar should be kept in the dark and brought out only for temporary display, even then, the specimens will slowly alter to orpiment.

Few miners are interested in arsenic minerals; in most

instances arsenic minerals present a problem to the separation of the ore. Few smelters want to buy ore that has a high arsenic content. For this reason most realgar is thrown out on the dumps of the mines if it can be easily separated from the ore. Realgar is common as a gangue mineral, so it is likely that mines may have specimens of realgar on their dumps. A search of mine dumps may turn up a few specimens, possibly crystals, along with other interesting minerals.

Excellent realgar specimens can be found at the Reward Mine near Black Diamond, 20 miles southeast of Seattle. Superb specimens of prismatic, bright red, realgar crystals on matrix have been recovered from this mine in recent years. The realgar occurs as single crystals and groups on a tan sedimentary rock and on coal. The crystals are attached to the walls of veins that are filled with calcite. Realgar is scattered through the veins, concentrated in places, almost missing in much of the veins.

The mine is located above the Green River across from the Green River Gorge Park (Fig. 90). The best realgar veins are located below the mine in the bank of the river. It is necessary to search the veins, breaking them up until a few red crystals are found. Then the remaining pieces of the vein around this spot are chiseled out and taken home. At home, the pieces of the vein are put in hydrochloric acid to etch away the calcite, hopefully exposing good crystals.

There is another report of realgar nearby. This is on the west bank of the Green River about 1 mile north of the Reward Mine. Realgar is reported to be concentrated in a dike on the side of a coal seam. It is possible that good crystals may occur at this deposit.

SIDERITE

Siderite is common in sedimentary rocks and as a gangue mineral in metallic mineral deposits. This iron car-

Fig. 89 *Realgar, Reward Mine. Largest crystal 3mm. Bob Jackson Collection.*

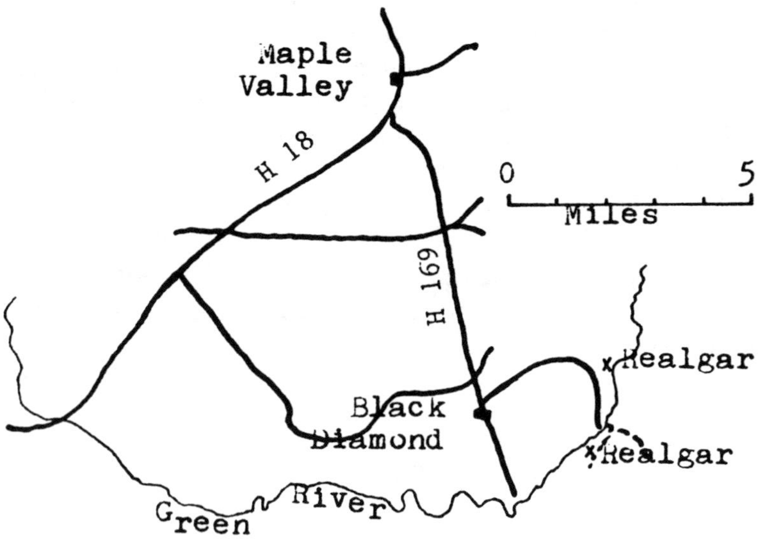

Fig. 90 *Black Diamond area*

168

bonate crystallizes in the hexagonal system; common habits are rhombohedral or spheroidal. The color is usually light to dark brown, yellowish-brown or greenish-brown. It is rarely clear or light yellow. The mineral is common in the metallic ore-producing districts of the state, but rarely in good crystal form. The best occurrence of siderite in Washington is the spheroidal groups found in the basalts of the eastern part of the state.

The area immediately east of downtown Spokane has produced the best and most specimens of spheroidal siderite. Specimens are found in outcrops, road cuts, and spoils piles in basalt vesicles which also contain micro crystals of hyalite pseudomorphs after anhydrite, crystobalite, feldspars, apatite and pyrrhotite. The spheres are commonly up to .5 inch in diameter; specimens over one inch have been reported. They are often covered by a botryoidal coating of hyalite. Groups are common; most spheres are attached to the vesicle wall, so that only about one half to five-eighths of the sphere has formed. A few spheres can be found that are nearly completely formed (Fig. 91).

Most of the spheres are a greenish-brown or olive color, but a few are reddish-brown or black. The reddish-brown spheres have a glassy luster or play of colors. The black spheres are usually altered to goethite.

All of the specimens are found in dark gray to black Columbia River basalt. A hammer is needed to break open the basalt chunks, a chisel may be useful to reduce the size of the matrix. Trimming must be done carefully, the siderite is often detached by the vibrations of the trimming process. Numerous specimens with seemingly empty vesicles should be retained by the collector for a careful search under a microscope for micro-size crystals.

A similar occurrence of siderite can be found in the basalts around Pullman. During a study of the minerals in basalt vesicles by the author while attending Washington State University, some 15 minerals were identified by X-ray and physical methods. Most of the minerals were micro-size, a few were visible only under a microscope.

Fig. 91 *Siderite, Spokane. Large sphere is .9 inch in diameter. Rudy Tschernich Collection.*

Fig. 92 *Siderite, Busby. Spheres have a glassy luster and are iridescent. Vug is .75 inch across.*

170

This suite of minerals include quartz, opal, cristobalite, labradorite, magnetite, pyrite, calcite, hyalite opal and siderite. A similar suite of minerals was reported by E.V. Shannon (1923) in Spokane.

Siderite is most abundant as spheroids, but not uncommon as rhombohedrons or bow-tie shaped specimens (Carter, C.L. 1974). Most of the latter two forms were less than .1 inch in length, but the spheres reached diameters of .25 inch. The spheres are similar to those at Spokane, but a greater amount of them are altered to goethite. Generally they are black when altered. A few of the siderite spheres have a play of colors on the surface, especially those from the road cut near Busby (Fig. 92).

These specimens can be found in most of the basalt outcrops and road cuts in and around Pullman. The author found good specimens in the road cut immediately south of Busby, the railroad cuts north of Reaney Park, and in the road cut at the turn off to the cemetery west of town (Fig. 93).

Fine micromounts of the minerals that occur with the siderite can be collected. Most of these are not described elsewhere in this book, so they are briefly discussed here. Pyrite was scarce as shiny cubes or "rosettes" of cubes that were less than .25 inch across. Magnetite was rare as octahedrons less than 1 mm across. These small crystals were found attached to acicular crystals of apatite or crystals of labradorite. Franklinite occurred in a few specimens found near Reaney Park as pseudohexagonal plates less than 2 mm in diameter. Calcite was scarce as massive vesicle fillings or small rhombohedral crystals. Cristobalite formed cube-octahedral shaped crystals less than 1 mm in diameter. Usually only hemispheres of this mineral formed in the vesicles. Some acicular crystals of an apatite group mineral, possibly wilkeite, were found in a few vesicles. Labradorite occurred as simple pinacoid or more complex tabular pinacoidal crystals.

Micromounters may find that collecting specimens of minerals in basalt vesicles may be rewarding. The above minerals were found in vesicles in only one area in Eastern

Washington. It is likely that these same minerals and others occur in numerous localities in the Columbia River basalts.

Siderite spheres similar to those found in Spokane and Pullman have been reported in the basalt in the Pasco area. The occurrence is the same as in Spokane. Any collectors hunting in this area should search the basalt vesicles for other interesting minerals.

A few crystals of siderite have been found at Denny Mountain. Brown massive siderite occurs with the specular hematite at the raspberry quartz scepter locality. A few vugs lined with clear siderite crystals can be found in the massive material. Most crystals are attached to the walls, and only partially formed, but a few good crystals do occur. Some of the crystals are phantoms; they have a yellow phantom inside of a colorless crystal. A silver, feathery, metallic mineral occurs in some of the vugs. This may be goethite. It is quite soft, very thin and easily crushed.

Occurrences of siderite can be found in remote areas along the Middle Fork of the Snoqualmie River. Such deposits may be found with quartz crystals in some of the breccia deposits. Some fine micromount to thumbnail specimens may be found by the fortunate collector in this area.

SILVER

Like gold, silver crystallizes in the isometric system. Crystals are usually malformed and in branching arborescent groups or in irregular masses, scales or wire. The native metal is soft with a hardness of 2½ to 3, malleable, and ductile. It has a high specific gravity, 10.5 when pure, so it is often found concentrated with gold in stream gravels.

The most pleasing specimens are those of branching arborescent groups or wires. These are found in vugs in ore deposits and are mostly supergene in origin. Specimens are common in the oxidized portions of such deposits. This environment provides excellent matrixes of other well crystallized secondary minerals. Silver is found associated with many other metallic minerals, but rarely in the same specimen. Calcite is a common matrix mineral. The calcite may be crystals or may be massive. It can be dissolved with hydrochloric acid to expose beautiful silver specimens.

Very little native silver has been collected in Washington. Traces have been reported from several copper, lead and gold mines. Some specimens of plates and scales on joint surfaces may be found on mine dumps. These often require a sharp eye to first notice them, as a careful search of freshly broken rock is required to collect such specimens.

A large pocket of wire silver was discovered in the IOU Mine northwest of Kettle Falls. It occurred as a large lens of native silver in a wide quartz vein in a granite dike. No other silver was reported at the mine. Other mines in the area reported minor amounts of silver, but there were no other large pods of silver recovered.

Small wires of native silver have been found in vugs in quartz veins in granodiorite and pyroxenite in the Covada Mining District. The district has produced silver, gold, lead and zinc. There are numerous mines with abundant dumps in which collectors may search for specimens. To find the small silver specimens it is necessary to break the pieces of quartz on the dump to expose small vugs. Some native gold may also be found in this district in the quartz. There are numerous mines and dumps in a large area west, north and east of Covada Lake. This is an old mining district that has experienced only minor exploration in recent years.

The Old Dominion Mine in the NW¼ of Sec. 9, T35N, R40E, 7.5 miles by road from Colville had produced silver, lead and zinc from galena, cerussite, sphalerite, and native silver. Secondary minerals are abundant near the surface, including wire silver. Sulfide ore bodies in limestone have been mined out. A few specimens have reportedly been found on the dump.

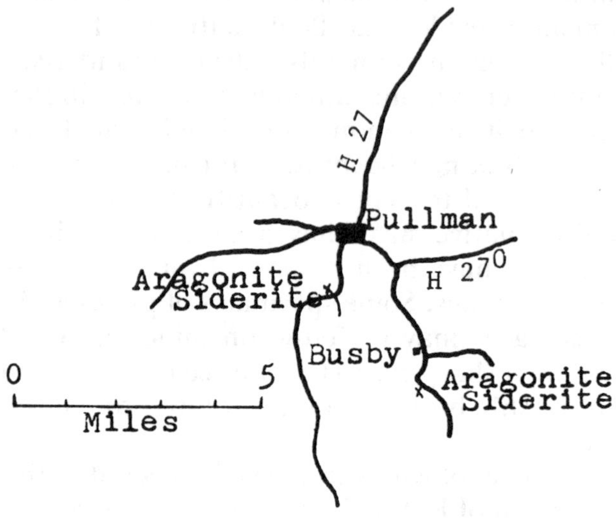

Fig. 93 *Pullman area*

Fig. 94 *Siderite with quartz, chlorite and epidote, Crawford Creek. Large crystal is .5 inch high.*

SMITHSONITE

Smithsonite is an uncommon mineral of zinc-bearing ore deposits. It is most abundant in ore deposits which have weathered in arid climates. This zinc carbonate crystallizes in the hexagonal system. Crystals are usually small and poorly developed rhombohedrons. It usually occurs as botryoidal crusts of blue to green color; these crusts make very attractive mineral specimens.

Light blue crusts occur at the Josephine Mine. The mine is located a short distance from the Z Canyon Road about 3 miles from Metalline Falls. The mine produced lead and zinc ore from a replacement deposit in the gray to black limestone of the Metaline formation. The mine is closed; small specimens have been found on the dump.

A similar occurrence is reported from the Oriole Mine nearby. Here it occurs with sphalerite, tetrahedrite, malachite, azurite and hemimorphite. Thick botryoidal crusts of a chestnut brown color are reported at the Black Rock Mine near Northport.

SPHALERITE

The zinc sulfide mineral, sphalerite, is found in several of the mining districts of Washington. It is usually associated with copper, lead and iron sulfides. Sphalerite crystallizes in the isometric system. Crystals are commonly in complex forms of tetrahedrons, dodecahedrons and cubes. The color is white to yellow, brown, red or black. The color darkens as the iron content increases. The names blackjack and rubyjack are applied to the black and red forms. In Canada and parts of the U.S. the black form is called marmesite.

Crystals are not common in this state. Most sphalerite is in masses or irregular grains. The mineral is abundant in mines in the carbonates of northeastern Washington, yet the vein deposits of the Cascades are probably the best sources of small crystals. Some veins and fracture fillings in the deposits contain small crystals often embedded in calcite. By etching out the calcite with hydrochloric acid small crystals may be exposed in the most unlikely vein. Such specimens are suitable for micromounts or rarely thumbnail specimens.

The Calhoun Mine in Stevens County has produced a few small crystals. These are found in small vugs with white tremolite. The sphalerite crystals are usually less than .13 inch across. Most of them are brown in color; they make fine micromount specimens.

Striated tetrahedrons to .5 inch occur with quartz, ankerite and chalcopyrite at the Mono Mine on the Miller River. The occurrence is in vugs in an altered volcanic breccia. Sphalerite and chalcopyrite are abundant as the filling between breccia fragments. Euhedral crystals are numerous in some vuggy zones.

At the Leta Mine, near North Bend, dark brown sphalerite crystals occur with bladed pyrrhotite and ankerite crystals. The sphalerites are striated and often twinned. They reach a diameter to .75 inch.

Dark brown crystals to .45 inch diameter can be found in vugs with tan ankerite crystals at the Cleveland Mine near Hunters. The occurrence is with mimetite, hemimorphite and others. Most specimens are found by searching the dumps at the mine.

STIBNITE

Stibnite is a common mineral in the mining districts of the state, and several mines have been developed to recover this iron antimony sulfide. Crystals are mostly slender prisms in radiating groups or bladed crystals of the ortho-rhombic system. It is silver gray in color; the shiny crystals make very fine specimens.

This mineral is usually found in quartz veins in granite, gneiss, limestone and shales. It is associated with intrusive rocks. Common mineral associates are other antimony sulfides, galena, cinnabar, sphalerite, barite, realgar and gold.

The most noted occurrence of stibnite crystals in the state is in the Lucky Knock Mine in Okanogan County, near Tonasket. There are two adits and four open pits at the mine. Stibnite crystals up to 8 inches long and .75 inch across (Fig. 96) have been found in the pit 150 feet west and 120 feet above the portal of the lower adit. The crystals were found in vugs in a fault that is exposed in the pit. Most of the crystals were partially altered to stibiconite or coated with limonite. Terminations were poor.

Underground workings have exposed large masses of stibnite, but no crystals have been found. The host rock is limestone, quartzite, schist and phyllite. There is very little or nothing of interest on the dump at this mine.

Small stibnite crystals have been found in the Pinnel prospect in Pend Oreille County. This stibnite deposit is located on the east bank of Skookum Creek eight miles east of Usk. The Kings Lake Road is followed for two miles from Usk and then a branch road is followed for another six miles. From here an old mining road can be followed by foot for a few hundred feet to the mine.

Mineralization is in Cambrian quartzite. Outcrops are scarce, the dump is small, and the mine is not safe. Small quartz veins can be found in the quartzite that is exposed in a few surface outcrops and two pits that are above the portal of the tunnel. There are a few vugs in the quartz veins that contain drusy quartz and rarely small stibnite crystals. The crystals are coarsely bladed or are fine ac-cicular crystals in radiating clusters, less than .4 inch. These specimens are of greater interest to micromount collectors.

A few crystals can be found in veins in Bear Basin on

the North Fork of the Snoqualmie River 45 miles east of Seattle. The basin is reached by taking the North Fork of the Snoqualmie River Road for 20 miles from North Bend to the Lennox Creek Road. Follow this road for about three miles to the trail to the basin. A hike of about two hours is required to reach the basin. The first hour is spent hiking up the switch backs of an old mining road, which may be maneuvered by four-wheel drive vehicles. Most of the numerous mines can be reached by a 15 minute hike from the cabin in the basin.

Excellent stibnite can be found on the dump of an adit that is northeast of the cabin. It is the next to last tunnel on the north end of the east wall. The dump is overgrown and the adit is in a steep gulch under a waterfall and is not easily seen from the cabin. There is a chockstone over the portal of the tunnel and a large pool of water at the portal.

This tunnel follows a vein of pyrite, arsenopyrite, stibnite and quartz. Most of the vein is highly weathered. Some good material may be found on the dump or in the vein where it is exposed on the steep cliff in the gulch above the portal of the tunnel. Climbing up the gulch is difficult, it should not be attempted without rope.

Most of the stibnite is massive, but there are a few small vugs. Small stibnite crystals are found in the vugs with small quartz crystals (Fig. 97). Crystals are generally less than .25 inch in length, but they have been reported up to .50 inch long. These shiny terminated crystals are best suited for micromounts. Similar crystals may occur in a few of the other mines in the basin, and some of the stibnite is altered to stibiconite.

Collectors will only need a hammer if collecting on the dump. The massive vein material has a dark brown weathered surface that should be chipped to be sure it is stibnite. Careful breaking of this massive material may yield small crystal-filled vugs. A hard hat and climbing equipment will make the collecting trip safer.

The host rock of the basin is mostly diorite of the Snoqualmie batholith. The two mines on the north side of the basin are in or along a large alaskite dike in the diorite. Blocks of diorite in the many talus slopes are often coated with black tourmaline (schorl) crystals or small quartz crystals.

Stibnite occurs as excellent crystals to one inch long at the Silverleaf Mine, Covada Mining District. Fine clusters with quartz crystals, valentinite, and cervantite can be collected (Cannon, 1975). Stibiconite forms alteration rims around some of the stibnite.

STRONTIANITE

Strontianite is a strontium carbonate. It most often occurs in sedimentary rocks, but can also be found as an accessory mineral in some metallic mineral deposits. It crystallizes in the orthohombic system and forms good accicular crystals, botryoidal masses or concretions. It is generally white or colorless, but may be light pink, green, yellow or brown.

A few fine specimens have been collected at the Alverson prospect on the southeastern tip of Fidalgo Island southwest of LaConner in Skagit County. Open pits expose veins of strontianite in a serpentinized dunite. A few good specimens of botryoidal strontianite have been collected in the pits and trenches (Fig. 98). Realgar, pyrite and arsenopyrite can be found in veins nearby, but are not euhedral crystals. Access is difficult by land, but the deposit can be reached by boat.

There are a few reports of strontianite in some of the mining districts of the state, most of which are probably of massive material. It is possible that a few euhedral crystals or botryoidal masses may be found in some of the mines or on their dumps.

Fig. 96 *Stibnite, Lucky Knock Mine. Specimen is 9.5 inches long and partially altered to stibiconite. Ore Inc. Collection.*

TOURMALINE

The tourmaline group contains three separate minerals, schorl, dravite and elbaite. Schorl, the sodium iron aluminum borosilicate hydroxide is probably the only tourmaline group mineral that is found in Washington. This black mineral usually occurs as prismatic crystals, sometimes in groups. It commonly occurs in pegmatitites, as coatings on joints in granitic rocks and rarely in metallic mineral deposits.

The few known occurrences in this state are mostly in igneous rocks and ore deposits. Most occurrences do not contain good euhedral crystals. In the St. Helens, Bear Basin and Silver Creek Mining Districts schorl forms radiating groups on joint surfaces in diorite and metasedimentary rocks. Some of the specimens may make fine cabinet material, and a few terminated crystals can be found in miarolitic cavities.

At the Mystery Mine in the Monte Cristo Mining District (see jamesonite for a description and access) excellent groups of schorl can be found on the mine dumps. Here the tourmaline forms clusters of radiating acicular crystals up to .75 inch in diameter. Several metallic minerals may be found on this and other dumps in the district. This is a very scenic, rugged, mountainous area.

A similar occurrence of schorl is at the Cleopatra Mine on the West Fork of the Miller River in King County. As at the Mystery Mine, the crystals form clusters of radiating acicular crystals which may be over 1.75 inches in length and 1.25 inches in diameter. There are a few additional mines with dumps in this area.

Schorl can be found on Eightmile Creek and Cannon Mountain southwest of Leavenworth. This locality is reached by taking the Icicle Creek Road to Eightmile Creek and following this road to the parking area. Pegmatites are exposed in logging roads to the left and to the south on Cannon Mountain. The country rock is the Chiwakum Schist.

The tourmalines are dark green to black and up to 4 inches long. Most commonly the crystals are frozen in quartz, but a few terminated crystals may be found with muscovite crystals in vugs. Hard rock collecting tools will be needed at this locality.

In the Sherman Pass area west of Kettle Falls dark green translucent crystals to .75 inch occur in pegmatites. Associated minerals include bright green microcline, smoky quartz, spessartine and muscovite. The quartz crystals reach lengths up to 2.5 inches and some show complex skeletal forms. Spessartine occurs as orange trapezohedrons to .5 inch.

WULFENITE

The tetragonal mineral wulfenite occurs in tabular square crystals or rarely pyramidal crystals. Its color is yellow, orange, red, gray or white, and its composition is lead molybdate. Most common in the arid southwestern part of the United States, it occurs in oxidized portions

Fig. 97   *Stibnite, Bear Basin. Crystals average 3mm long.*

of lead mines.

Yellow, orange and brown crystals occur at the Avondale Dome Mine north of Colville (Cannon, 1975). The crystals are tabular to pyramidal and up to .25 inch across. They occur on coxcomb hemimorphite associated with galena and cerrusite. Specimens are not abundant, but by carefully searching the mine dump a few specimens can be found.

ZEOLITES

There are thirty-three species of minerals that belong to the zeolite group. Of these, about ten of them are com-

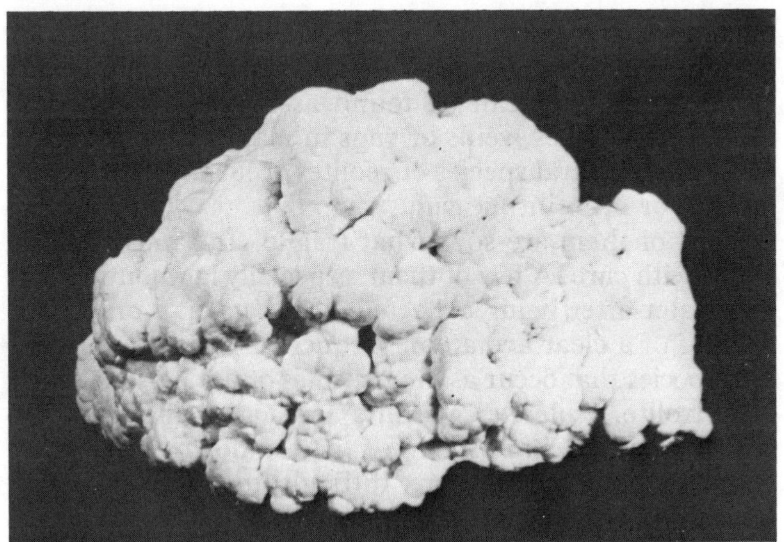

Fig. 98   *Strontianite, Fidalgo Island, 4.4 inches. Rudy Tschernich Collection.*

mon, these include analcime, stilbite, natrolite, heulandite, thomsonite, laumontite, mordenite, chabazite, scolecite and mesolite. The remaining minerals vary from uncommon to quite rare, some of them occurring at only one or two known localities.

The five most common minerals of the group are listed and described below. Analcime crystallizes in the isometric system. Crystals are usually trapezohedrons and are clear to white in color. Natrolite crystallizes in the monoclinic system. The crystals are usually pseudo-orthorhombic, acicular prisms, clear to white in color. Chabazite belongs to the hexagonal crystal group; most of the crystals are rhombic with nearly cubic angles. Heulandite and stilbite are monoclinic. Heulandite crystals are usually tabular, the centers of the crystals are commonly wider than the ends, giving the crystals a coffin shape. Stilbite crystals are tabular, and are usually in sheaflike groups. Heulandite is often colorless, but stilbite is most often white in color.

Zeolites are most common as vesicle fillings or vein fillings in volcanic rocks. They are secondary in origin, and are often associated with calcite, quartz, prehnite and apophyllite. They can be found as masses or euhedral crystals in vesicles, veins or vugs in many volcanic rocks. Most often several species of zeolites occur together in one locality or even in one vug.

Most of them are somewhat fragile, and must be collected with care. A few of them, especially laumontite, will lose water after being collected. If not given a protective coating of a clear acetate or lacquer they may fall apart. The species that occur as long needles or are hairlike, such as natrolite, scolecite, mesolite and mordenite, are very susceptible to being broken or matted down. They must not be touched, and are difficult to protect during transport.

Collecting these minerals is often difficult, because of the fragility and occurrence in hard rock. Some collectors remove them with a large piece of matrix attached, then carefully clean and trim them at home where more time and care can be given. Quite often good specimens are easily detached from the rock matrix during collecting.

Zeolites are dissolved by acids, so they are difficult to clean if dirty. They are often coated with mud and clay that has been washed into the vugs by water that has seeped down the numerous joints in the volcanic rocks. The mud can usually be removed by soaking the specimen in detergent or sometimes bleach. The long hairlike or needle-like crystals of natolite, scolecite, or mesolite are difficult to clean. If the dirt is not stuck on the crystals but merely lying amongst the crystals, then it can be removed with a needle. Hold the specimen upside down and carefully pick out the grains of dirt. Most of these minerals of acicular habit will mat down if put in water to clean them. If they are coated with mud or clay that can not be picked out, then immerse one poor quality specimen in water to see if it is safe to do so. If the crystals do not mat and stick together then carefully immerse and clean each specimen individually in detergent. Cleaning zeolites is often a long and tedious process.

there are numerous localities in western Washington where good quality zeolite specimens can be collected. Only a few localities have been reported in eastern Washington. On the west side of the state, zeolites are found in most of the large areas of Eocene age volcanic rocks. Many of these basalts, andesites and breccias are vesicular or highly jointed. They were good hosts to the secondary fluids that deposited zeolites and other minerals.

In eastern Washington the zeolites occur in the Columbia River basalt. There are abundant microcrystals of minerals in the vesicles of the basalts in some areas, but zeolites are uncommon. As crystal collecting becomes more popular there may be more localities discovered, there is a good potential for zeolites in the thousands of square miles of basalt on the east side of the state.

One of the best discoveries of zeolites was at the Skookumchuck Dam near Tenino. The town of Bucoda is also close to the dam, and specimens are often labeled with Bucoda as the locality. During construction of the dam, which started in 1969, numerous pockets and vugs lined with zeolites, quartz, apophyllite and calcite were discovered in the basalts. The basalts are Eocene age flows of the Northcraft formation. At the dam site the basalts are gray to green in color, and highly vesicular in some places and not in others.

The earth-fill dam was constructed to supply water to the Centralia Coal Electric Plant in the Hanaford Valley south of Tenino. The dam is on private land and access is not allowed. No collecting can be done at this locality.

Cavities in the basalt were discovered up to 3 feet by 4 feet by 8 feet in size. Large stilbite, mesolite and analcime crystals were found in these, while the numerous small vesicles yielded fine specimens of stilbite, heulandite, calcite and quartz (Tschernich, 1972).

The following minerals, listed in their paragenetic sequence (Tschernich, 1972), were found at the dam site, calcite, heulandite, mordenite, quartz, okenite, apophyllite, stilbite, mesolite, analcime, thomsonite, laumontite and chabazite. Of these calcite, heulandite, quartz, stilbite and mordenite were the most abundant.

One of the most spectacular minerals at this locality was the large acicular crystals of mesolite. This mineral usually occurs as fine hair-like crystals, but at this locality crystals were found that were up to 3.5 inches long and .25 inch in diameter. They occurred as individual clusters on stilbite or as fine hair-like sprays on thomsonite balls (Figs. 99 and 100).

Stilbite was found as small doubly terminated individual crystals and groups of crystals. They reached a length of over 4 inches. Loose doubly terminated crystals were found free in the mud inside of some of the large cavities. Small stilbite crystals were often found as the matrix for later mesolite, analcime and heulandite (Figs. 101 and 102).

Analcime was found as colorless to white trapezohedral crystals. They were common in sizes around .4 inch in diameter and a few were found that reached about 1.75 inches in diameter (Fig. 102). They formed good specimens as singles or groups with other zeolites.

Heulandite occurred as sharply terminated coffin-shaped crystals. They were white to pink in color and found in combination with other minerals. Mordenite was usually white and mostly found as "matted hair." Quartz was quite abundant in several habits, it commonly lines the cavities with only the termination showing, usually only three of the rhombohedral faces. It occurred as micro-sized spheres resembling fish eggs, or grains of rice in other pockets.

Okenite was found as chalky, white hemispheres commonly with prismatic crystals of apophyllite. Thomsonite was found as white hemispheres upon which mesolite grew and as small individual crystals. Laumontite was found as white prismatic crystals. Rhombohedral chabazite crystals were not abundant.

Good zeolite specimens were discovered during construction of Mossyrock Dam, southwest of Morton. The minerals are found in Eocene basalts similar to those at the Skookumchuck locality. Fine crystals of heulandite, stilbite, apophyllite and quartz were collected, but little material is available today. Basalt outcrops in the vicinity of the dam occasionally produce specimens.

East of Woodland, along Canyon Creek and Tum Tum Mountain fine specimens of zeolites including large laumontite crystals have been collected (Fig. 104). Good scolecite specimens have been collected at Elk Mountain, northwest of Canyon Creek (Fig. 105). This locality also produces apophyllite, levyne, and stilbite. North of here, along the Toutle River, some fine stilbite specimens have been recovered. These localities are shown on Fig. 106. Most of the land in this area is privately owned by various timber companies or is part of the Gifford Pinchot National Forest. Logging roads in the area are generally open spring and fall, but not during the snow or fire season.

Good natrolite and analcime specimens have been collected from a quarry near the end of the Lincoln Creek Road west of Centralia. The quarry is accessible by a dirt road from the end of the Lincoln Creek Road (Fig. 107). There is a gate at the junction of the two roads, but it is usually open. The land is owned by Weyerhaeuser Timber Company.

The host rock of the deposit is black Eocene age basalt. The lower part of the quarry is columnar basalt, but the top 30 feet is highly fractured and appears to be water-laid (Fig. 108). Zeolites line the numerous fractures. Near the top, most of the crystals are weathered, they are white and chalky. Farther down in the basalt they are fresh, the analcime and natrolite are often water-clear when fresh.

Analcime coats the joints with crystals up to .4 inch across. Crystals of natrolite radiate from some of the analcime up to 1 inch long and .1 inch in cross section (Fig. 109). With hard work, fine specimens of these minerals can be recovered.

The fractured nature of the basalt makes collecting these specimens not too difficult. A small pick, chisel or bar can be used to loosen the pieces of basalt and remove the zeolites. Extreme care must be taken because the natrolite clusters are fragile.

Several zeolites can be collected on Rock Candy Moun-

Fig. 99 *Mesolite, Skookumchuck Dam, 2.4 inches across. Rudy Tschernich Collection.*

tain which can be reached by taking State Route 8 (the Ocean Beach Highway) from Interstate 5 at Tumwater and following it for 10.5 miles to the Summit Lake exit. Instead of turning north to the lake, turn south and follow the road under the power lines for 1.7 miles. Turn left and follow this road for .3 miles to the road cut where the crystals are found.

The zeolites are found in vesicles in basalts in the road cuts on both sides of the road. Hard rock collecting tools are needed, but the crystals are fragile. Minerals found include scolecite, stilbite, levyne, calcite and possibly heulandite and mordenite. Volcanic rocks are extensive in this area, so there may be other zeolite occurrences nearby.

Small analcime, calcite and mesolite specimens may be

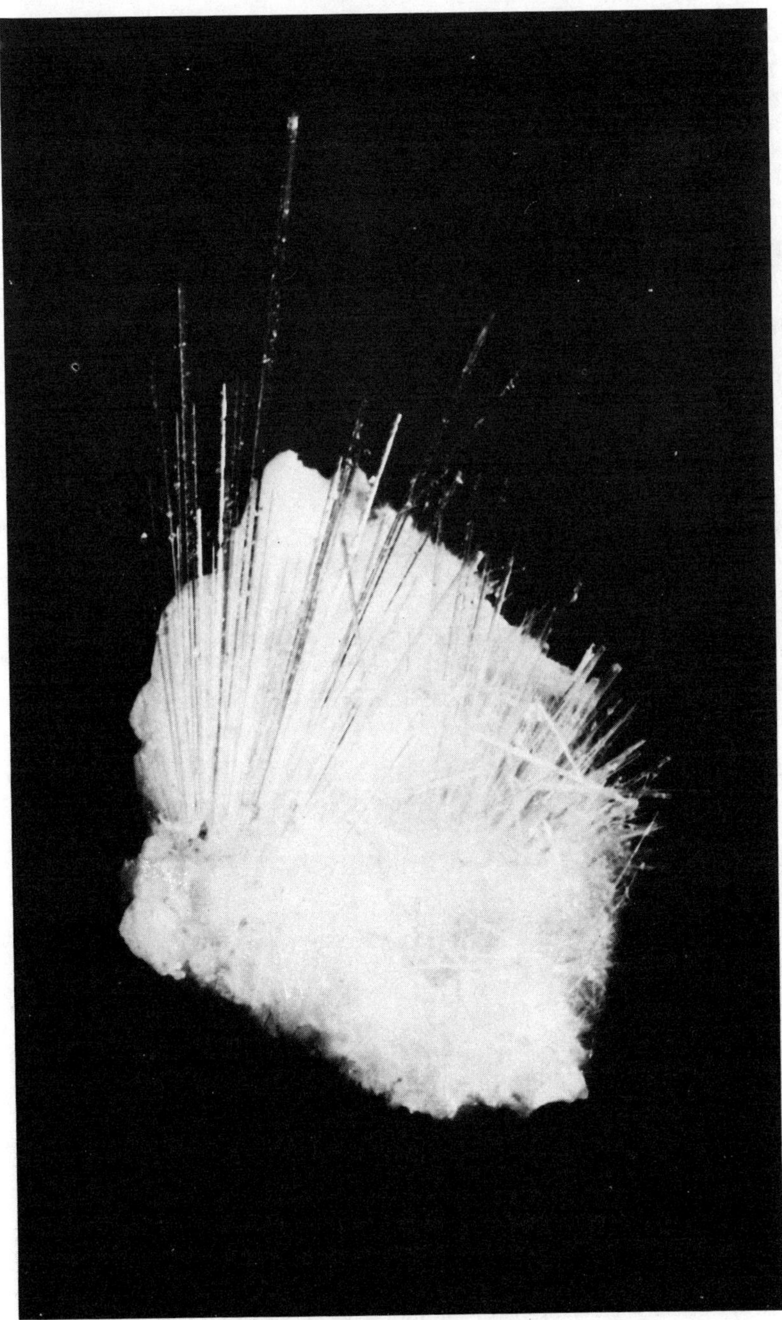

Fig. 100   *Mesolite, Skookumchuck Dam. .75 inch high.*

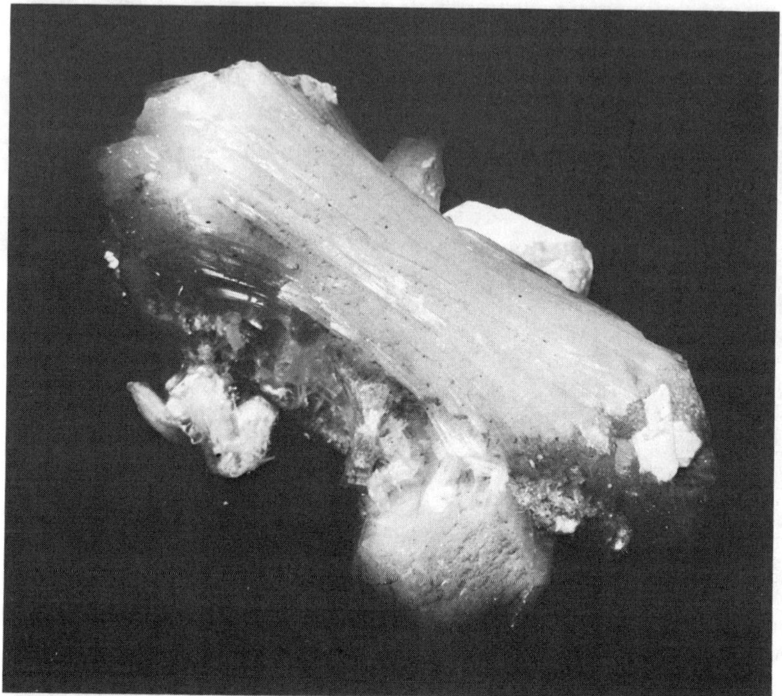

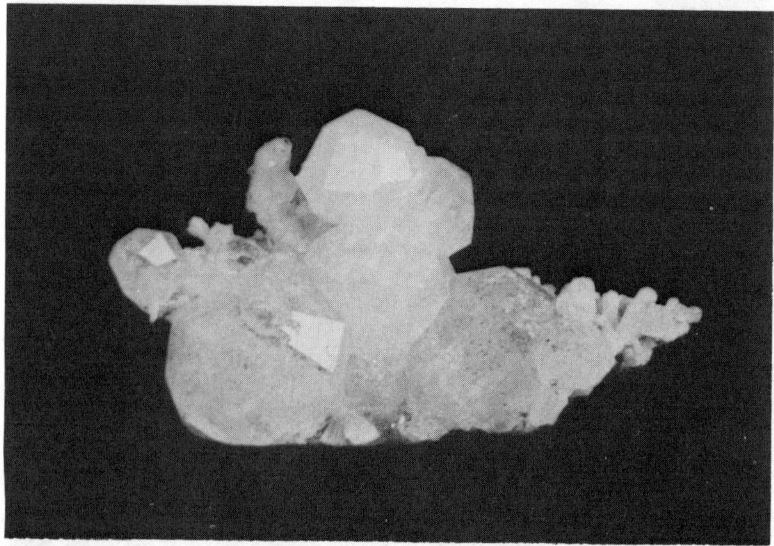

Fig. 101   *Stilbite, Skookumchuck Dam. 2.75 inch long. Rudy Tschernich Collection.*

Fig. 102   *Analcime, stilbite, Skookumchuck Dam, 3.5 inch across. Rudy Tschernich Collection.*

190

Fig. 103 *Analcime, Skookumchuck Dam, 2.4 inches across. Rudy Tschernich Collection.*

Fig. 104 *Laumontite, Canyon Creek. Large crystal is 1 inch high. Rudy Tschernich Collection.*

Fig. 105 *Scolecite, Elk Mountain, 2.2 inches across.*
Rudy Tschernich Collection.

collected from a quarry just northwest of Oakville. the quarry practically undercuts the west side of State Route 8, and can be reached by taking the third left turn when heading north from Oakville.

Fine specimens of acicular scolecite crystals were found in a quarry at Pe Ell. The slender white crystals radiate inward from the walls of vesicles in a gray basalt, (Fig. 110). The crystals are found in lengths up to 1.75 inches. At this time the quarry is not being worked, so collecting is strictly hard work.

Excellent specimens of analcime, apophyllite, stilbite, scolecite, laumontite and calcite can be collected at the Pioneer Quarry south of Bremerton. The analcime occurs in crystals to .4 inch in diameter and apophyllite occurs as colorless, white and pale green crystals to 1.5 inches. The stilbite crystals are up to 1 inch long and of white, tan and pale orange colors. Scolecite occurs in crystals up to 3 inches long. Laumontite reaches a maximum length of 1 inch.

This occurrence is in vesicles in dark gray basalt. Collecting requires hard rock tools and hard work.

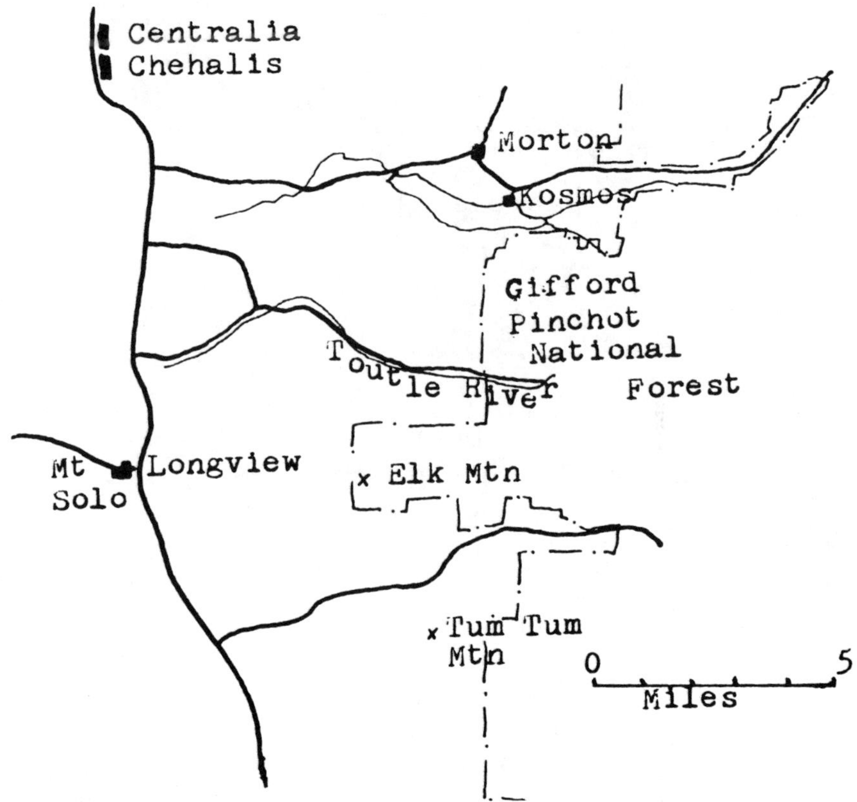

Fig. 106 *Lewis and Cowlitz counties area*

Stilbite crystals to 1 inch long were collected from a miarolitic cavity on Lennox Creek near the North Fork of the Snoqualmie River. The occurrence is about 300 yards upstream from the first bridge over Lennox Creek after the turnoff from the North Fork Road. This cavity lies on the north side of the road partially concealed in the ditch.

Quartz and natrolite crystals were recovered with the stilbite. The occurrence is in a dark gray contact phase of the Snoqualmie batholith. Searches of the area have failed to reveal any other crystal filled cavities. There is an unusual pyrite occurrence across the creek, north of

193

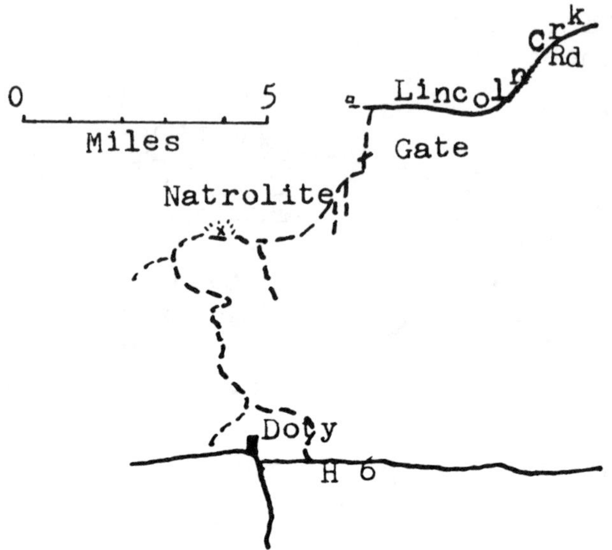

Fig. 107   *Doty area*

Fig. 108   *Lincoln Creek Quarry. (Kathy Ream photo)*

the bridge near this cavity. Small crystals of pyrite occur in books of biotite in a coarse grained granite. Specimens are interesting but not very showy.

Analcime and natrolite can be found in vesicles in basalt on Mount Solo on the west edge of Longview. The natrolite is scarce, but the analcime is abundant as trapezohedral crystals under .25 inch in diameter. The crystals are clear and make excellent micromounts and miniatures.

Heulandite, mordenite, analcime, chabazite and phillipsite can be found in the basalt outcrops around the town of Kalama. During the construction of Interstate 5 good specimens were collected; most of them were of small size.

Excellent specimens of pink mordenite can be collected along Rock Creek northwest of Stevenson in the Columbia River Gorge. The mordenite occurs as hair-like crystals coating vesicles in a weathered basalt. Vesicles are numerous, but most of them are entirely filled with mordenite so there are no terminations. Terminated crystals can be found in vesicles that reach dimensions of 1 to 2 inches.

Clinoptilolite, dachiardite, ferrierite, mordenite and stilbite occur in roadcuts around Altoona. The clinoptilolite forms crystals to .13 inch and the dachiardite occurs as microscopic sheaf-like clusters. Ferrierite occurs as small radial clusters under .13 inch in diameter. These crystals occur in vesicles in basalt.

A very rare mineral, paulingite, was discovered in basalt at Rock Island Dam south of Wenatchee. This was the first discovery of this mineral. Since then one locality has been discovered in Idaho. The paulingite is found in rounded pieces of basalt in the river bottom. Much of this basalt was dredged up during construction of the dam. There is some disagreement over the origin of the basalt. Kamb and Oke (1960) believe that the rounded pieces of basalt originated some distance upstream. Several collectors in the state believe that the pieces are nearly in place, and are only rounded as much as is expected from a fractured, vesicular basalt on the bottom of the river.

Rhombic dodecahedrons of paulingite are attached to

Fig. 109   *Natrolite, Lincoln Creek, .75 inch high.*

Fig. 110   *Scolecite, Pe Ell. Vug is 1 inch in diameter. Rudy Tschernich Collection.*

the walls of the vesicles. They appear as only hemispheres, .1 to 1 mm in diameter. Filiform pyrite occurs with the paulingite, as does erionite. The pyrite forms filaments that are .05 mm in square cross section. They are up to 4 mm in length, some of them make right angle bends. A few cubes have been reported. The erionite forms fine transparent acicular crystals. They occur in isolated singles and as radiating clusters. They are up to 2 mm in length. Heulandite, phillipsite and calcite are found in the same rocks.

The paulingite and pyrite is scarce, and very little of the basalt pieces can be found now. The rare paulingite and unusual filiform pyrite make excellent micromount specimens.

The author found one specimen of chabazite in the basalts of the Snake River Canyon downstream from Clarkston. The specimen was found on the canyon wall above the stile over the fence near the Moses opal deposit. Crystals are less than 3 mm across and are white in color. There may be other occurrences of zeolites in the basalts of this canyon and elsewhere in eastern Washington.

There are other reports of zeolites in Washington. Scattered occurrences of dachiaride, erionite, and ferrierite have been reported. A search of the basalt outcrops in western and eastern Washington may reveal additional fine zeolite localities. See update.

UPDATE

Since Gems and Minerals of Washington was last printed, several significant discoveries have been made. New exposures within broad areas of mineralization have increased Washington's specimen production. The hobby has experienced an influx of young, active collectors, who are adding to the list of minerals found in the state. Roadbuilding on forest lands continues to expose new collectable materials, and to provide access to formerly inaccessible areas. This recent history of discovery leads one to believe that many potential new finds await the active collector in Washington. The following additions are arranged alphabetically by type, just as in the original text.

GEMS

Agate: Road construction along the I-5 corridor between Chehalis and Portland has exposed a number of areas of agate-bearing soil. Nodules found have ranged from carnelian to blue-gray. Such soil disruption is usually temporary, and localities are covered as quickly as new ones are exposed. Construction areas in the characteristic clay-rich soil of decomposing basalt should be checked frequently.

Moss agate, sagenitic agate, jaspers, and thundereggs are being found on Weyerhaeuser land in the Enumclaw area. From Hwy. 410 enter the 6100 road about 9 miles east of Mud Mountain Dam. Drive south 7.4 miles to road 6110; turn right for 1.2 miles. Hunt thundereggs in the rock slide. A swampy area 2.6 miles further along road 6100 has produced agate nodules to 60 pounds. The roads south and east of here are excellent areas to prospect for gem material. Two areas in the cliffs north of road 6100 have produced small amounts of precious opal. Access is by hiking from road 6100 in the vicinity of the 9-mile gate, or from the first road to the north past the 9-mile gate. The Weyerhaeuser Company publishes excellent free maps to their lands. See "Useful Addresses" on page 215.

Amber: A land trade on Tiger Mountain east of Issaquah has provided access to the long-closed locality on 15 Mile Creek. From Route 18 take the west side logging road up over the ridge and 3 miles down the west side of the mountain to the bridge over 15 Mile Creek. Hike 2 miles up the creek. The amber occurs as nodules in shale and sandstone along the walls of 15 Mile Canyon. Carbonized wood and leaf fragments serve as indicators in the sediments. Amber nodules average ½ inch in size, but pieces to 4 inches do occur.

Obsidian: Obsidian-like glass has been found in road cuts on Weyerhaeuser road 6015-4. Access is via the 6100 road, then 1 mile on 6015-4. Pieces are small. Cabochons cut from the dark gray to black glass often show metallic-appearing iridescences.

Rhodonite: An excellent deposit of rhodonite veins near Split Rock in Skagit County produces bright pink material. The deposit is located between two small lakes on the north side of private forest road SWB 1540, 8/10 mile west of the junction with SWB 1500. These roads are accessed from the Lake Cavanaugh road which intersects highway 9 near Big Lake. Beware of logging trucks!

MINERALS

Andalusite: Altered andalusite crystals to 1½ inches were found in a shear zone on the Quartz Creek claims, west of the Taylor River in King County. Non-commercial collecting is permitted on the Quartz Creek claims.

Epidote: Epidote crystals to 2 inches, mostly imbedded in quartz seams, are being found in a contact zone located just east of the summit of Green Mountain, a few miles east of North Bend. Further north along the ridge that extends from Green Mountain to Bessemer Mountain, a hematitic zone in an old road cut produced hundreds of excellent pseudomorphs of quartz crystal-coated hematite after epidote. Many of the original epidotes were twinned. Though difficult to clean, when the iron oxides are removed single pseudomorphs up to 10 inches long emerge.

Feldspars: Sharp crystals of adularia occur in a road cut on Weyerhaeuser land just south of Lake Hancock in King County. In 1981 several chloritized quartz-carbonate veins were cut by road work less than ½ mile south of the lake. These produce attractive chlorite phantoms in small quartz crystals as well as groups of adularia to 2 by 2 inches. Further east, new road work in 1983-84 uncovered similar material. Another Weyerhaeuser road in this district exposed aplite dikes with cavities of albite, smoky quartz, and black tourmaline. The tourmaline is of acicular habit; individual crystals reach 1¼ inches. Small but striking specimens of gemmy smoky quartz on lustrous white albite were abundant when the road was cut in 1981. Collectors have exposed more aplite vugs in the cliffs east of the cut. Remnant boulders from the roadbuilding have also been productive. Access is via the main trunk road south and west of Lake Hancock, then by a secondary road that runs east between Lake Moolack and SMC Lake. The productive cut is on the east side of the road after the road turns north around the end of the lake. On your way out, notice the attractive purplish contact breccia where the granite rock contacts the old metamorphic series rocks.

Garnet: Gemmy almandine garnets are found on Ruby Mountain along the North Cascades Highway (Route 20). The highly modified dodecahedrons often exhibit so many crystal faces that they appear almost round. The mica schists exposed on the rugged face of the mountain reportedly have produced ½ inch garnets, though the average size encountered in talus below and along Ruby Creek is considerably smaller.

Jamesonite: A productive area for small jamesonite crystals in calcite-filled sulfide veins has been found in the Silver Creek area north of Index. Access is from Route 2. Follow the river road north; cross the river at the bridge to the abandoned townsite of Galena. Proceed up the logging road (usually washed out) to a creek on the west side, opposite Big Gulch on the east side. You'll need a topographic map to find Big Gulch. Hike up the west creek about ½ mile. At this point, the veins cross the creek running southwest-northeast. Remove entire pieces of vein, taking them home to dissolve the calcite in acid. Other crystalline minerals in the veins include sphalerite, acanthite, arsenopyrite, boulangerite, pyrrhotite, and pyrite.

Pyrite: Rounded, complex crystals of pyrite have been produced from a breccia within the Alpine Lakes Wilderness area. Located on a steep hillside along the ridgetop separating the Middle Fork of the Snoqualmie River from Burntboot Creek, the breccia has vugs to 12 inches with pyrites to 2 inches. Collectors MUST contact the US Forest

Service North Bend office for permission to collect in the wilderness.

Road cuts in the Metalline Falls area of Pend Orielle Co. produce filiform pyrite crystals embedded in mountain leather.

Quartz: A commercial specimen-producing operation on Denny Mountain in 1980 and 1981 produced many fine specimens of amethyst scepters. At least one vug large enough to stand upright in was encountered. White overgrowth scepters (non-amethystine) are quite abundant in the talus. The adventuresome collector who comes equipped with rock climbing hardware and heavy tools may still produce a white scepter plate or two, occasionally with the added attraction of a hematite rose. The prospect of a collector with hand tools uncovering an amethyst scepter vug remains, as before this latest operation, bleak. The occasional amethyst scepter is still available to those who screen the hematite-quartz talus. Because the cirque bottom was almost completely cleaned of pre-existing talus by the specimen miners, the best pickings for screeners are in the canyon below the cirque.

Small stout amethyst crystals resembling the Brazilian material have been recovered from a canyon on the east side of the Middle Fork of the Teanaway River north of Cle Elum. The canyon is reached by a 2-mile hike from the Teanaway River road.

Large Japan Law twins have been collected from a contact area in the Hansen Creek drainage. Attractive manganite balls and chlorite occur as phantoms in a large percentage of single quartz crystals to 3 inches. The locality is cut near the end of a logging spur road about ½ mile due south of the amethyst scepter location. Road access is along the trunk road, past the micro anatase locality (see Update: Rare Minerals), then left onto the spur.

Clear quartz crystals to 5 inches and occasional amethyst scepters have been recovered from several thick quartz veins outcropping on a prominent ridge north of the Snoqualmie Lake Potholes. The area is reached by trail from Lake Dorothy. Similar quartz veins have been observed on high ridges north of Deer and Bear Lakes, also reached via the Lake Dorothy Trail. Dark purple amethyst crystals to 6 inches, associated with larger, milky quartz crystals have been found near the Ferry/Okanogan County line, west of Curlew Lake. The ridgetop deposits are above old mine workings. Near Republic, the Sheridan Mine has produced small quartz crystals heavily included with purple fluorite, which appear amethystine.

Rare Minerals: Several rare minerals, two of them newly described, are being collected from the alkali-rich granites of the Golden Horn batholith, near Liberty Bell Mountain. The area is reached on Route 20, the North Cascades Highway. Okanoganite, zektzerite and gagarinite, complex silicates of rare earth metals, crystallize in miarolytic cavities in the granite. Associated minerals include smoky quartz, zircon, riebeckite-arfvedsonite, fluorite, bastnaesite, elpidite, acmite, astrophyllite, allanite, prehnite, titanite, thorite, gadolinite, and euxenite. While most of these minerals are micros, zekzerites to 1 inch have been found. Zektzerites are pink to yellow to white, and some specimens are gemmy enough to facet. Okanoganite is rhombohedral, but commonly forms fourling twins that are pseudotetrahedral. The largest found so far are about 1/6 inch. The most productive areas have large exciting-looking cavities which are apparently weathering features, unrelated to crystallization. Local collectors humorously term these "Wookie Tubes." Several as yet unidentified minerals have been collected, and the possibility of other rare mineral discoveries is high. Collecting in areas nearest the road has been heavy. A several day trip is recommended for those wishing good chances at recovering several of the rare mineral suite. The alkali granite is most exposed above the tree line, making this one of the most scenic collecting areas in the state.

Anatase has been found in two places near Snoqualmie Pass. Lustrous small crystals on quartz crystals occur in sporadic vugs on Silver Peak, south of Snoqualmie Pass. The easiest hiking route is from the ridge on which the

Large pyrite crystal, 5 inches across, held by collector. Spruce Ridge.

ski areas operate south of the highway. Brilliant micro crystals can be collected associated with apatite and quartz at a cut on the Hansen Creek Road. Using the map on page 157, bear right instead of turning left to the quartz locality, and continue about ½ mile to the cut.

Zeolites: There have been several discoveries of new zeolite localities and new zeolite minerals in the northwest, including Washington.

The Robertson Quarry near Shelton has produced excellent analcime crystals to 2 inches. Heavy tools are required to break basalt pillows to access the pockets. The quarry is worked infrequently, and collectors are not admitted when machinery is present.

Ferrierite occurs in the Columbia River basalts around Altoona. The source is a quarry on the east end of town, but most collecting has been done in the riprap along the

Japan Law twin quartz from Hansen Creek. Largest twin "ear" measures 4¼ inches.

Columbia River. Ferrierite forms colorless or white blades and coatings in vesicles and fracture fillings in basalt breccia. Associates include calcium-clinoptilolite, dachiardite, mordenite, calcite and chalcedony. (Wise and Tschernich, 1976).

Large specimens of stilbite and mesolite can be collected from vuggy areas in road cuts in the Toutle River area, particularly in higher areas accessed by the South Fork Toutle Rd.

Pale pink zektzerite on smoky quartz from the Golden Horn batholith. Zektzerite is ¼ inch. Bob Jackson collection.

"Crown of Thorns": an interesting quartz and pyrite specimen produced by the collapse of a vug ceiling into a pyrite plate below and continued precipitation of quartz and pyrite to "heal" the two pieces together. Major specimens such as this are often named by the collectors. Bob Jackson collection.

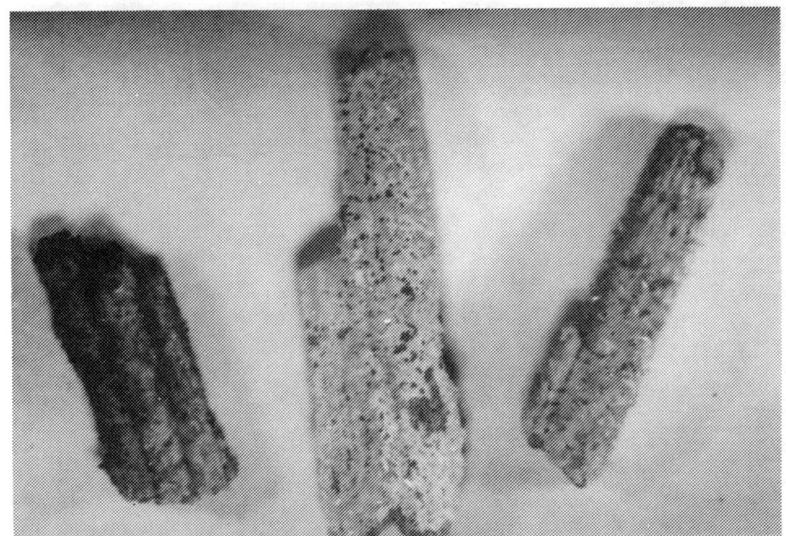

Hematite pseudomorphs after epidote, Bessemer Ridge. Specimens are partially to fully coated with lustrous small quartz crystals. Largest crystal, 4 inches.

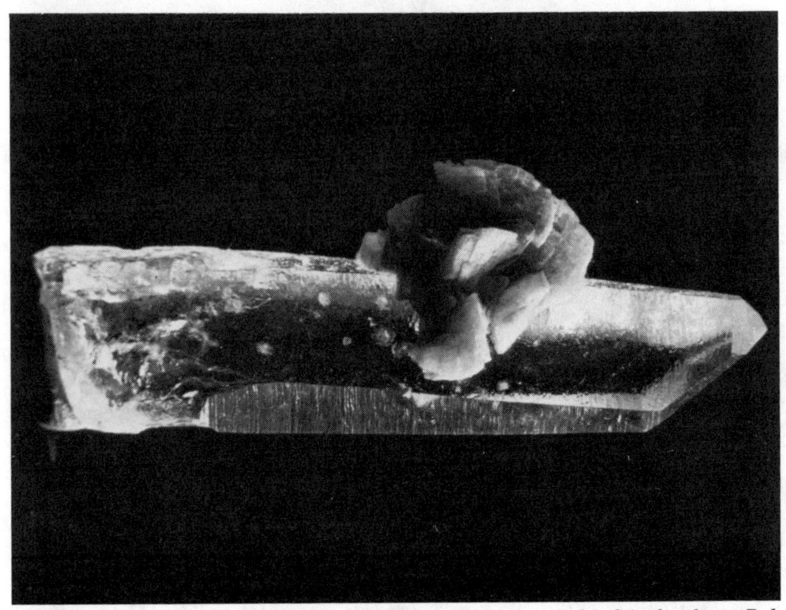

Ankerite "rose" on clear quartz, Spruce Ridge. Quartz crystal is 3 inches long. Bob Jackson collection.

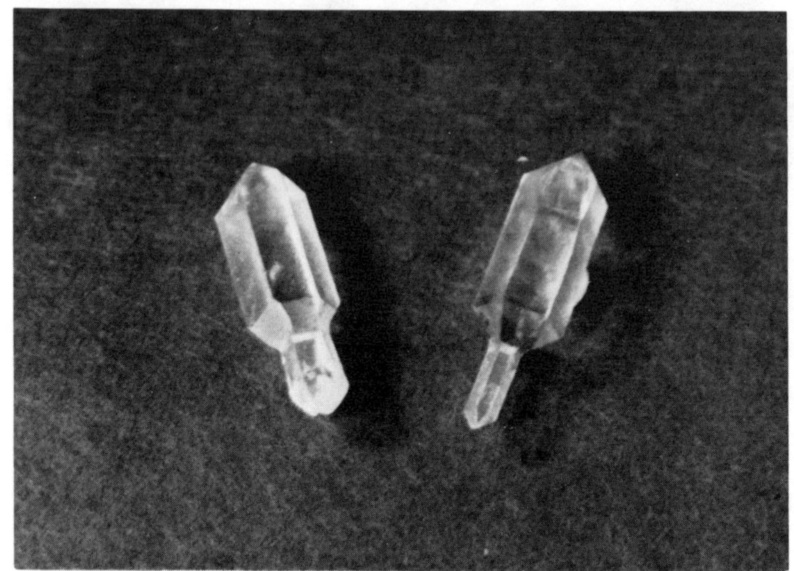

Sceptered quartz from Spruce Ridge. This locality produces thousands of clear dramatic quartz scepters every year for jewelry and esoteric uses.

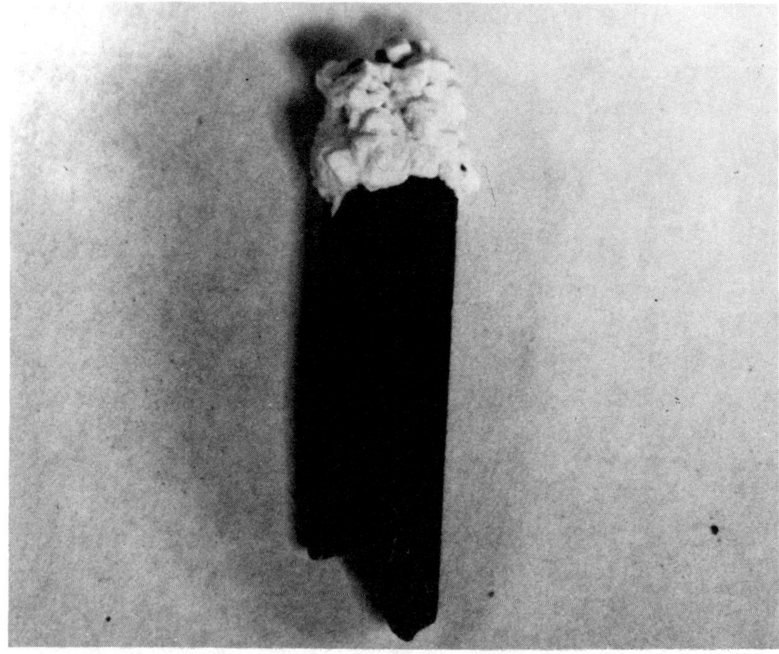

Two inch Riebeckite/Arfvedsonite. Termination coated by albite. Golden Horn batholith.

Sceptered Quartz and Pyrite. Spruce Ridge.

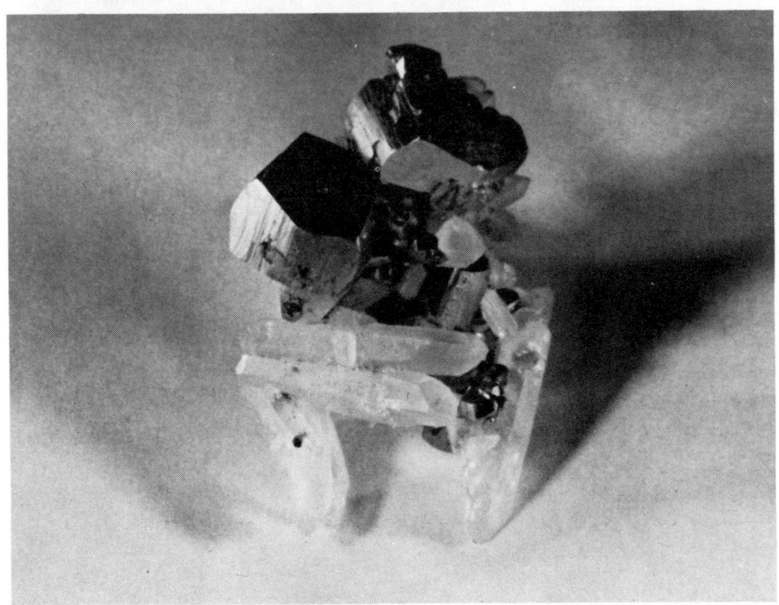

Floater of doubly-terminated quartz and pyrite crystals, Spruce Ridge. Specimen measures 3 inches. Collecting trips are offered on this claim — see "Useful Addresses."

209

Quartz and Tabular Pyrite. Spruce Ridge.

BIBLIOGRAPHY

Amstutz, G.C., 1958, Coprolites, a review of the literature and a study of specimens from southern Washington: Jour. of Sed. Pet., V .28, No. 4, p. 498-508.

Bennet, W.A.G., 1962, Saline Lake deposits in Washington: Wash. State Div. of Geol. and Earth Resources, Bull. 49, 129p.

Boggs, R.C., 1980, Okanoganite, a new rare-earth borofluorosilicate from the Golden Horn batholith, Okanogan County, Washington: Am. Min. V. 65, p. 1138-1142.

Brockett, Ron, 1974, The Moscow opal mines, 1890 to 1893: 63p.

Cannon, Bart, 1975, Minerals of Washington: Cordilleran, 184p.

Carter, C.L., 1974. Varieties of siderite occurring in vesicular basalt between Moscow, Idaho and Pullman, Washington: The Min. Rec., V.5, No. 4, p. 157-195.

Clevinger, W.R., 1970 Chehalis-Newaukum carnelian agate collector's guide: Seattle University Bookstore, 29p.

Drake, H.C., 1956, Northwest Gem Trails: Mineralogist Pub. Co., 80p.

Dunn, P.J., Rouse, R.C., Cannon, B., Nelen, J.A., 1977, Zektzerite: a new lithium sodium zirconium silicate related to tuhualite and the osumilite group: Am. Min. V. 62, p. 416-420.

Foraker, David, 1972, Collecting Ellensburg's rare agates and petrified wood: Gems and Minerals, no. 422, p. 46-47, 65-66, November.

Hadley, W.D., 1975, Walker Valley Amethyst: Rock and Gems, August 1975, p. 56-61, 77.

__________ , **1976, The Wonderful miniatures of Walker Valley: Rock and Gem, February 1976, p. 56-59.**

Huntting, M.T., 1949, Perlite and other volcanic glass in Washington: Washington Div. of Geology and Earth Resources, Rep. of Inv. No. 17, 77p.

________________ , **1960, Inventory of Washington minerals, Part I nonmetallic minerals, Part II metallic minerals: Washington Div. of Geol. and Earth Res., Bull. 37.**

Jackson, Bob and Jackson, Kay, 1974, The Rockhounds guide to Washington, V. I, 48 p.

________________ , **1975, The Rockhounds guide to Washington, V. II, 45p.**

Kamb, W.B. and Oke, W.C., 1960 Paulingite a new zeolite in association with erionite and filiform pyrite: Am. Min. V. 45, No. 102, p. 79-91.

Kenoyer, Pat, 1974, Yakima Canyon wood: Gems and Minerals, No. 436, p. 20-22.

Lapidary Journal, 1965, Washington field trips near Cle Elum: V. 19, No. 5, p. 108-609.

Livingston, V.E., 1959, Fossils in Washington: Washington Div. of Geol. and Earth Res., Inf. Circ. No. 33, 35p.

__________ , **1971,** Geology and mineral resources of King County, Washington: Washington Div. of Geol. and Earth Res., Bull. 63.

Lucas, J.C., 1974, The actinolite crystals of Wenatchee Lake: Lap J., V. 27, No. 11, p. 1654, 1656, 1658-1659, 1668-1669, February.

Mills, J.W., 1960, Geologic setting of the nickel occurrence on Jumbo Mountain, Washington: Min. Eng., V. 12, No. 13.

________________ , and others, 1971, Bedded barite deposits of Stevens County, Washington: Ec. Geol., V. 66, p. 1157-1163.

Moen, W.S., 1964, Barite in Washington: Washington Div. of Geol. and Earth Res., Bull. 51, 112p.

__________ , 1969, Geology and mineral resources of Whatcom County, Washington: Washington Div. of Geol. and Earth Res., Bull. 57.

Parsons, G.C., 1962, Practical gem knowledge for the amateur: Lap. J., Inc., 140p.

Palache, C., and others, 1944, 1951, 1962, Dana's system of mineralogy, V. I, 1944, V. II, 1951, V. III, 1962, John Wiley and Sons, New York.

Pattee, E.C. and others, 1968, Beryllium resources of Idaho, Washington and Oregon: U.S. Bur. of Mines, Rep. of Inv. No. 7148, 169p.

Purdy, C.P., Jr., 1951, Antimony occurrences of Washington: Wash. Div. of Geol. and Earth Res., Bull. 39.

__________ , 1954, Molybdenum occurrences of Washington: Wash Div. of Geol. and Earth Res., Inf. Circ. 18.

Ream, L.R., 1974, Washington Gem Jade: Lap., V. 28, No. 4, p. 708-711.

__________ , 1975, Nephrite in Washington: Lap. J., V. 29, No. 9, p. 1748-1757.

__________ , 1977, Denny Mountain Quartz: Min. Rec., V. 8, No. 1.

Shannon E.V., 1923, On siderite and associated minerals from the Columbia River Basalts at Spokane, Washington: Proceedings of the U.S. Nat. Mus., V. 62, Art. 12, 19p.

Sinkankas, John, 1959, Gemstones of North America: Van Nostrand Reinhold Co., 675 p.

Snavely, P.D., and Wagner, H.C., 1963, Tertiary geologic history of western Oregon and Washington: Washington Div. of Geol. and Earth Res., Rep. of Inv. No. 22, 25p.

Tschernich, R.W., 1972, Zeolites from Skookumchuck Dam, Washington: Min. Rec. V. 2, No. 1, p. 30-34.

Weaver, C.E., Paleontology of the marine Tertiary formation of Oregon and Washington: Univ. of Wash. Pub. in Geol., V. 5, 803p.

Williams, J.R., 1974, Beachcombing in Jefferson County, Washington: Gems and Min., No. 435, p. 22-23.

Wise, W.S. and Tschernich, R.W., 1975, Cowlesite a new Ca-zeolite: Am. Min. V. 60, p. 951-956.

Wise, W.S. and Tschernich, R.W., 1976, Chemical composition of ferrierite: Am. Min. V. 61, p. 60-66.

Useful Addresses

Explorations Unlimited
Box 2652
Renton, WA 98056
Geologist/naturalist Bob Jackson offers collecting at the Spruce Ridge quartz and pyrite crystal locality and other Northwest areas. He also teaches adult and children's classes in popular geology. Brochure.

Washington Department of Geology
Department of Natural Resources
Olympia, WA 98504

Free pamphlet discusses collecting in Washington. Newsletter on more technical aspects of state geology. Publications lists. Excellent library. Mineral identification service.

Washington State Mineral Council
c/o Bob Pattie
4316 NE 10th
Renton, WA 98056

Organization to promote/preserve collecting localities in Washington. Newsletter.

Weyerhaeuser Company
33633 32nd Dr. S
Auburn, WA 98002

Free maps showing access to company-owned lands.

OTHER TITLES IN NORTHWEST/WASHINGTON GEOLOGY FROM JACKSON MOUNTAIN PRESS

The Rockhounds' Guide to Washington, by Bob Jackson. Three volumes. These frequently updated guides take collectors to a variety of gem, mineral, and fossil sites in Washington. Each volume contains different localities, 20 to 24 each. Over 15,000 copies in print. $3.95/volume. Volume II is out of print.

The Panners' Guide to Northwest Gold, by Bob Jackson. The best-selling guide to gold seeking in the Northwest. Covers the geology of gold; how to use the natural processes which concentrate gold in placer deposits to your advantage; the historic occurrences of gold; maps to producing, non-claimed areas; equipment; etc. $3.95.

Northwest Volcanoes: A Roadside Geologic Guide, by Lanny Ream. New in 1984, this book takes the traveller on the back roads, trails, and highways around the major volcanoes of the Cascade range. Lanny Ream describes the forces that shaped the Cascades in non-technical terms, and brings geologic history alive with guided stops at rock formations that play major roles in that history. $6.95.

St. Helens! The 1980 Eruptions, by Lanny Ream and Bob Jackson. The first book published following the historic eruptions. Explains the events leading to St. Helens renewed activity in words and pictures. $2.95.

Gems and Minerals of Washington, by Lanny Ream. A collector's reference to mineralogy and localities, by a well-known Northwest geologist. 220 pages. $7.95.

A special acknowledgement is offered here to Bart Cannon for the contributions his book, "Minerals of Washington," have made to the mineral section of the present work. Cannon's book is available from Cordilleran, 1041 N.E. 100 Street, Seattle, WA 98125. ($6.75 postpaid)